NELSON

VICmaths

VCE UNITS 3 + 4

specialist mathematics 12

mastery workbook

Greg Neal
Sue Garner
George Dimitriadis
Stephen Swift

Nelson VICmaths Specialist Mathematics 12 Mastery Workbook
1st Edition
Greg Neal
Sue Garner
George Dimitriadis
Stephen Swift
ISBN 9780170464079

Publisher: Dirk Strasser
Additional content created by: ansrsource
Project editor: Alan Stewart
Series cover design: Leigh Ashforth (Watershed Art & Design)
Series text design: Rina Gargano (Alba Design)
Series designer: Nikita Bansal
Production controller: Karen Young
Typeset by: MPS Limited

Acknowledgements
TI-Nspire: Images used with permission by Texas Instruments, Inc
Casio ClassPad: Shriro Australia Pty. Ltd.

ISBN 978 0 17 046407 9

Cengage Learning Australia
Level 7, 80 Dorcas Street
South Melbourne, Victoria Australia 3205

Cengage Learning New Zealand
Unit 4B Rosedale Office Park
331 Rosedale Road, Albany, North Shore 0632, NZ

For learning solutions, visit **cengage.com.au**

Printed in China by 1010 Printing International Limited.
1 2 3 4 5 6 7 26 25 24 23 22

Contents

Integration 133

Areas and Volumes of integration 164

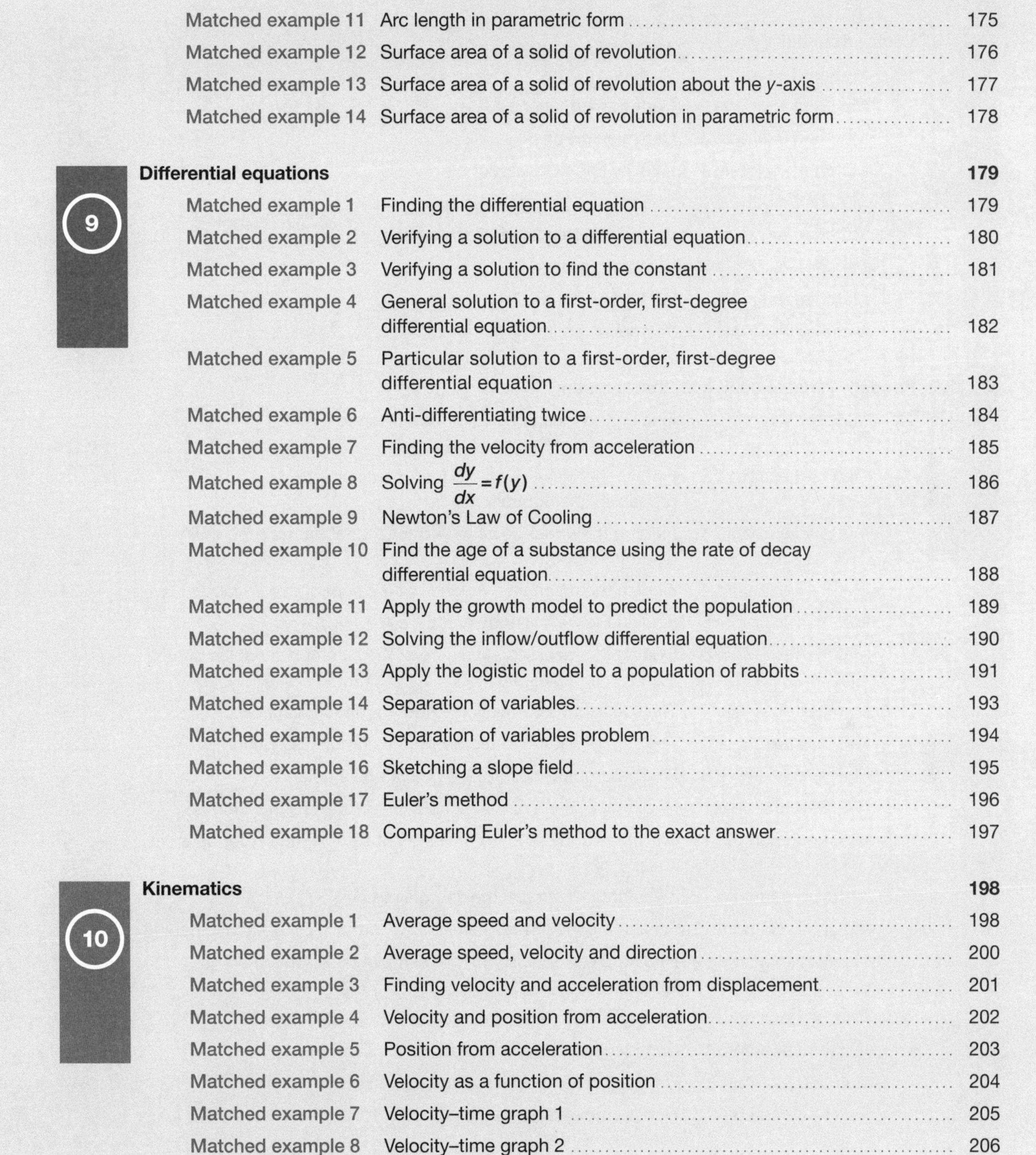

To the student

Nelson VICmaths is your best friend when it comes to studying Specialist Mathematics in Year 12. It has been written to help you maximise your learning and success this year. Every explanation, every exam hack and every worked example has been written with the exams in mind.

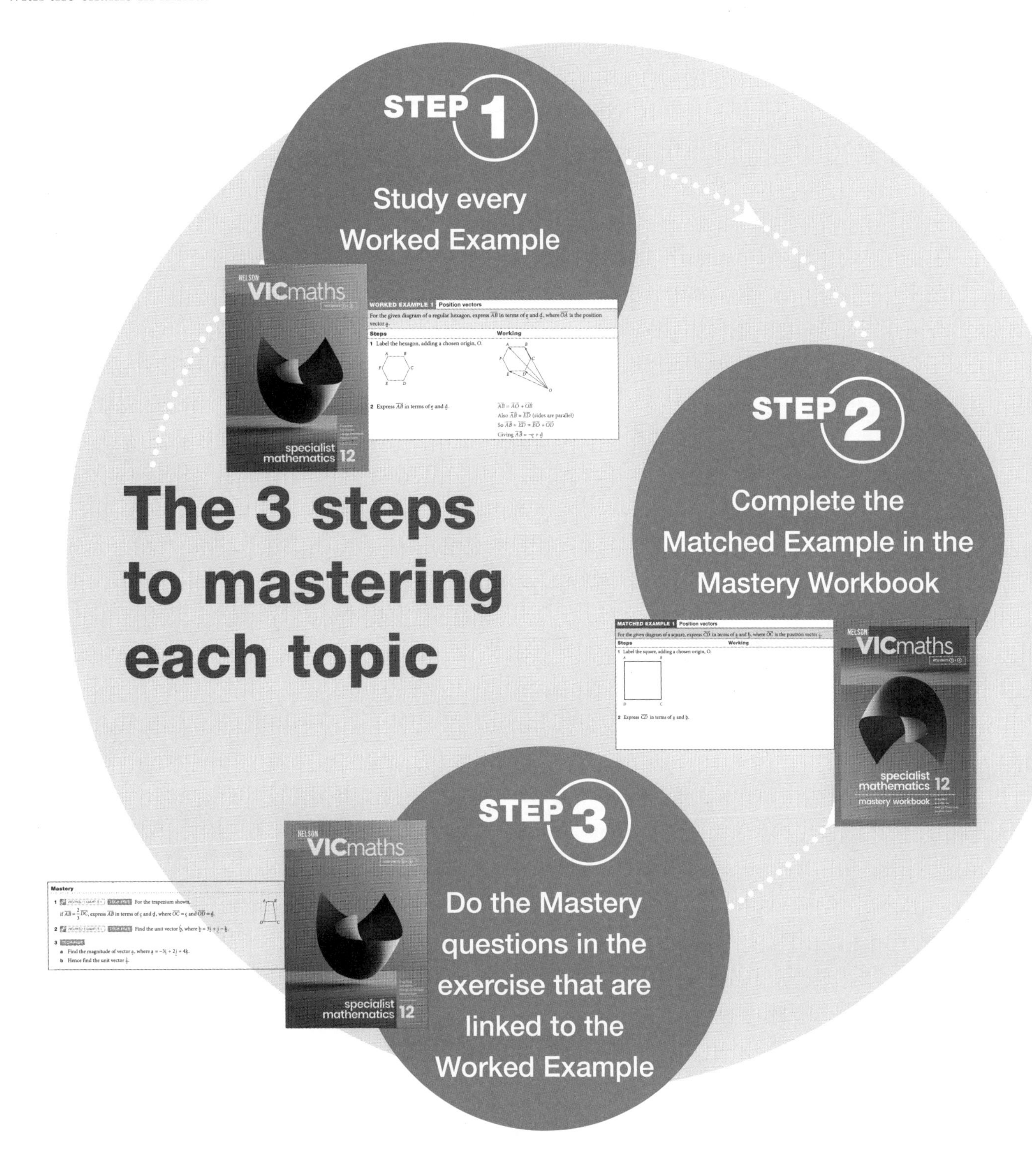

VECTORS

CHAPTER 1

MATCHED EXAMPLE 1	Position vectors
For the given diagram of a square, express $\overrightarrow{CD}$ in terms of $\underset{\sim}{a}$ and $\underset{\sim}{b}$, where $\overrightarrow{OC}$ is the position vector $\underset{\sim}{c}$.	
Steps	**Working**
1 Label the square, adding a chosen origin, O. 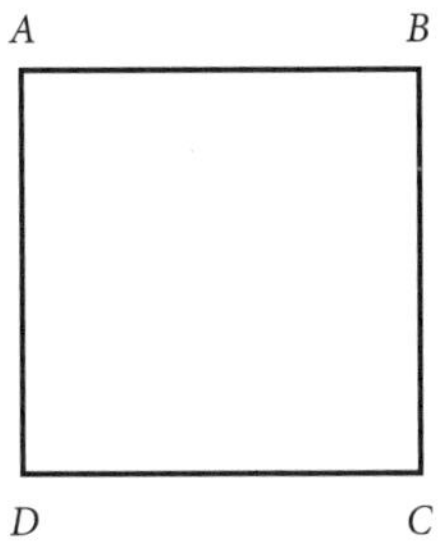	
2 Express $\overrightarrow{CD}$ in terms of $\underset{\sim}{a}$ and $\underset{\sim}{b}$.	

SB
p. 4

SB
Using CAS 1:
Operations with vectors
p. 5

p. 6

MATCHED EXAMPLE 2 Unit vectors

Find the unit vector $\hat{\underset{\sim}{b}}$, where $\underset{\sim}{b} = 2\underset{\sim}{i} - \underset{\sim}{j} - \underset{\sim}{k}$.

Steps	Working
1 Find the magnitude of $\underset{\sim}{b}$.	
2 Write down $\hat{\underset{\sim}{b}}$ using the formula $\hat{\underset{\sim}{b}} = \dfrac{\underset{\sim}{b}}{\lvert\underset{\sim}{b}\rvert}$.	
TI-Nspire	**ClassPad**

p. 8

MATCHED EXAMPLE 3 Linear dependence

Consider the set of four vectors $\underset{\sim}{a} = 2\underset{\sim}{i} - \underset{\sim}{j} + 3\underset{\sim}{k}$, $\underset{\sim}{b} = 4\underset{\sim}{i} - 2\underset{\sim}{j} + 6\underset{\sim}{k}$, $\underset{\sim}{c} = \underset{\sim}{i} - 3\underset{\sim}{j} + \underset{\sim}{k}$ and $\underset{\sim}{d} = 3\underset{\sim}{i} + \underset{\sim}{j} + 2\underset{\sim}{k}$.

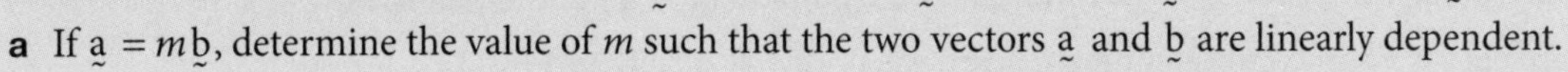

a If $\underset{\sim}{a} = m\underset{\sim}{b}$, determine the value of m such that the two vectors $\underset{\sim}{a}$ and $\underset{\sim}{b}$ are linearly dependent.

b Determine whether the three vectors $\underset{\sim}{a}$, $\underset{\sim}{c}$ and $\underset{\sim}{d}$ are linearly dependent.

Steps	Working
a Set up the statement $\underset{\sim}{a} = m\underset{\sim}{b}$, where $\underset{\sim}{a} = \begin{bmatrix} 2 \\ -1 \\ 3 \end{bmatrix}$ and $\underset{\sim}{b} = \begin{bmatrix} 4 \\ -2 \\ 6 \end{bmatrix}$. In this case, we will use the square bracket representation.	
b **1** Set up the statement $\underset{\sim}{a} = m\underset{\sim}{c} + n\underset{\sim}{d}$, where $\underset{\sim}{a} = \begin{bmatrix} 2 \\ -1 \\ 3 \end{bmatrix}$, $\underset{\sim}{c} = \begin{bmatrix} 1 \\ -3 \\ 1 \end{bmatrix}$ and $\underset{\sim}{d} = \begin{bmatrix} 3 \\ 1 \\ 2 \end{bmatrix}$.	
2 Equate $\underset{\sim}{i}$, $\underset{\sim}{j}$ and $\underset{\sim}{k}$ terms. In this case, we will use a table to solve equations.	
3 Use the components of $\underset{\sim}{j}$ and $\underset{\sim}{k}$ to solve equations for m and n. Substitute the values found for m and n and see if they hold for the components of $\underset{\sim}{i}$. Solutions do not hold for all three equations: linearly independent.	

p. 10

MATCHED EXAMPLE 4	Resolving vectors
a Write the vector $\overrightarrow{PQ}$ in component form given the points $P(1, 0, 2)$ and $Q(-1, 3, 3)$. **b** Write the vector $\overrightarrow{RS}$ in component form given the points $S(2, 2, 2)$ and $R(-3, 2, 1)$. **c** Write a vector of magnitude 5 units at an angle of 45° to the x-axis in 2D component form.	

Steps	**Working**
a Writes the points $P(1, 0, 2)$ and $Q(-1, 3, 3)$ using $\underset{\sim}{i}$, $\underset{\sim}{j}$ and $\underset{\sim}{k}$.	
b Use the distance and direction between each component of the points $S(2, 2, 2)$ and $R(-3, 2, 1)$ to write in component form.	
c Draw a diagram using the information given.	

MATCHED EXAMPLE 5 Angle between vectors and axes

SB p. 11

Find the angle between the vector $\underset{\sim}{a} = -2\underset{\sim}{i} - 3\underset{\sim}{j} + \underset{\sim}{k}$ and the z-axis. Write your answer in degrees to two decimal places.

Steps	Working
1 Identify a_3 and $\lvert\underset{\sim}{a}\rvert$.	
2 Use the formula $\cos\gamma = \dfrac{a_3}{\lvert\underset{\sim}{a}\rvert}$, where γ is the angle between the vector and the z-axis.	
TI-Nspire	ClassPad

p. 14

MATCHED EXAMPLE 6	Scalar product
a Find the scalar product of the vectors $\underset{\sim}{l} = 2\underset{\sim}{i} + 3\underset{\sim}{j} + \underset{\sim}{k}$ and $\underset{\sim}{m} = \underset{\sim}{i} - 2\underset{\sim}{j} + 2\underset{\sim}{k}$. **b** Find the value of x if the vectors $\underset{\sim}{l} = 2\underset{\sim}{i} - 3\underset{\sim}{j} + 5\underset{\sim}{k}$ and $\underset{\sim}{m} = 4\underset{\sim}{i} + x\underset{\sim}{j} + 2\underset{\sim}{k}$ are perpendicular.	
Steps	**Working**
a **1** Use the formula $\underset{\sim}{l} \cdot \underset{\sim}{m} = l_1m_1 + l_2m_2 + l_3m_3$.	
b **1** Use the formula $\underset{\sim}{l} \cdot \underset{\sim}{m} = l_1m_1 + l_2m_2 + l_3m_3$. **2** Use the formula for perpendicular vectors $\underset{\sim}{l} \cdot \underset{\sim}{m} = 0$ to solve for x.	

MATCHED EXAMPLE 7 Angle between vectors

Find the angle between the vectors $\underset{\sim}{l} = 2\underset{\sim}{i} + 3\underset{\sim}{j} - \underset{\sim}{k}$ and $\underset{\sim}{m} = 2\underset{\sim}{i} - 2\underset{\sim}{j} + 3\underset{\sim}{k}$ expressed in degrees correct to two decimal places.

Steps	Working
1 Use the formula $\underset{\sim}{l} \bullet \underset{\sim}{m} = l_1m_1 + l_2m_2 + l_3m_3$ to evaluate $\underset{\sim}{l} \bullet \underset{\sim}{m}$.	
2 Evaluate $\lvert\underset{\sim}{l}\rvert\lvert\underset{\sim}{m}\rvert$. 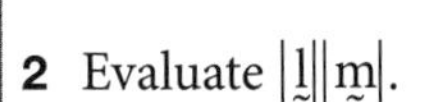	
3 Use the formula $\underset{\sim}{l} \bullet \underset{\sim}{m} = \lvert\underset{\sim}{l}\rvert\lvert\underset{\sim}{m}\rvert \cos\theta$ to find θ.	

TI-Nspire

ClassPad

SB p. 14

SB Using CAS 2: Scalar product and angle between vectors p. 15

p. 16

MATCHED EXAMPLE 8	Vector projections
Find the vector projections of $\underset{\sim}{a} = \underset{\sim}{i} + 4\underset{\sim}{j} + 2\underset{\sim}{k}$ parallel and perpendicular to $\underset{\sim}{b} = \underset{\sim}{i} - \underset{\sim}{j} - 2\underset{\sim}{k}$.	
Steps	**Working**
1 Find $\hat{\underset{\sim}{b}} = \frac{\underset{\sim}{b}}{\lvert\underset{\sim}{b}\rvert}$.	
2 Calculate the vector projection of $\underset{\sim}{a}$ on $\underset{\sim}{b}$.	
3 Find the vector projection of $\underset{\sim}{a}$ perpendicular to $\underset{\sim}{b}$.	

MATCHED EXAMPLE 9	Vector product
Find the vector product of the vectors $\underset{\sim}{a} = 5\underset{\sim}{i} + 3\underset{\sim}{j} - 2\underset{\sim}{k}$ and $\underset{\sim}{b} = 4\underset{\sim}{i} + \underset{\sim}{k}$.	
Steps	**Working**
1 Identify the coefficients in both vectors.	
2 Apply the vector product formula.	
3 Simplify.	

p. 20

Using CAS 3: Vector product p. 20

p. 21

MATCHED EXAMPLE 10 Vector product in determinant form

Find the vector product of $\underset{\sim}{a} = (4, -4, 2)$ and $\underset{\sim}{b} = (5, -6, 3)$.

Steps	Working
1 Identify the coefficients in both vectors.	
2 Apply the vector product determinant formula.	
3 Simplify.	

MATCHED EXAMPLE 11 Parallel vectors

p. 23

Consider the vectors $\underset{\sim}{a} = \underset{\sim}{i} - 2\underset{\sim}{j} - 2\underset{\sim}{k}$ and $\underset{\sim}{b} = \underset{\sim}{i} + \underset{\sim}{j} - \underset{\sim}{k}$.

a Find a unit vector that is parallel, and in the opposite direction, to $\underset{\sim}{a}$.

b Find the vector that is parallel to, and in the same direction as $\underset{\sim}{a}$ and with the same magnitude as vector $\underset{\sim}{b}$.

Steps	Working
a **1** Find the magnitude of $\underset{\sim}{a}$.	
2 Find the vector in an opposite direction to $\underset{\sim}{a}$.	
3 We know $\lvert\underset{\sim}{a}\rvert = 3$.	
b **1** Find the magnitude of $\underset{\sim}{b}$.	
2 Find $\hat{\underset{\sim}{a}}$.	
3 Multiply $\hat{\underset{\sim}{a}}$ by the magnitude of vector $\underset{\sim}{b}$.	

p.24

MATCHED EXAMPLE 12	Perpendicular vectors
Show that $\underset{\sim}{a} = 6\underset{\sim}{i} + 2\underset{\sim}{j} - 8\underset{\sim}{k}$ and $\underset{\sim}{b} = 4\underset{\sim}{i} - 4\underset{\sim}{j} + 2\underset{\sim}{k}$ are perpendicular.	
Steps	**Working**
1 Find the scalar product.	
2 State conclusion.	

MATCHED EXAMPLE 13 Ratios of vectors

SB p. 24

Consider the vectors $\overrightarrow{OA} = 2\underset{\sim}{i} - 3\underset{\sim}{j}$ and $\overrightarrow{OB} = \frac{1}{4}\underset{\sim}{i} + \underset{\sim}{j}$. Find the coordinates of M such that M is the point of trisection, closest to A, of the vector $\overrightarrow{AB}$.

Steps	Working
1 Find $\overrightarrow{AB}$.	
2 Find $\overrightarrow{AM}$.	
3 Find $\overrightarrow{OM}$.	
4 State the coordinates of M.	

p. 27

MATCHED EXAMPLE 14 Vector proof 1

In the square shown,
Which one of the following statements is true?

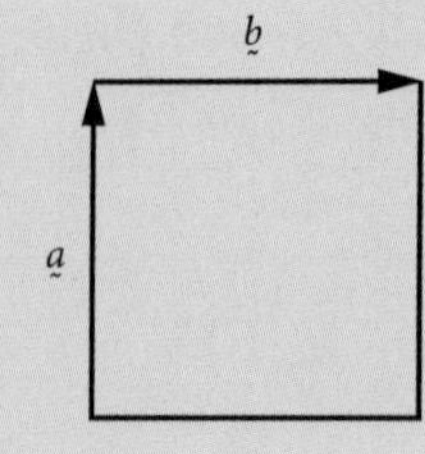

A $\underset{\sim}{a} = \underset{\sim}{b}$

B $\underset{\sim}{a} + \underset{\sim}{b} = 2\underset{\sim}{b}$

C $\underset{\sim}{a} \cdot \underset{\sim}{b} = |\underset{\sim}{a}||\underset{\sim}{b}|$

D $|\underset{\sim}{a}| = -|\underset{\sim}{b}|$

E $\underset{\sim}{a} \cdot \underset{\sim}{b} = 0$

Steps	Working
1 Consider what is given.	
2 Consider what is clearly wrong.	
3 Consider what we know.	
4 Consider what we know.	

MATCHED EXAMPLE 15	Vector proof 2
Use vectors to prove that the altitudes of a triangle are concurrent.	
Steps	**Working**
1 Sketch a diagram and set up the proof.	
2 State what needs to be proved.	
3 Consider $\overrightarrow{OA}$ is perpendicular to $\overrightarrow{BC}$. Express $\overrightarrow{OA}$ and $\overrightarrow{BC}$ in terms of $\underset{\sim}{a}$, $\underset{\sim}{b}$ and $\underset{\sim}{c}$.	
4 Consider $\overrightarrow{OB}$ is perpendicular to $\overrightarrow{CA}$. Express $\overrightarrow{OB}$ and $\overrightarrow{CA}$ in terms of $\underset{\sim}{a}$, $\underset{\sim}{b}$ and $\underset{\sim}{c}$.	
5 Add the equations $\overrightarrow{OA} \cdot \overrightarrow{BC} = 0$ and $\overrightarrow{OB} \cdot \overrightarrow{CA} = 0$	
6 Express $\underset{\sim}{a}$, $\underset{\sim}{b}$ and $\underset{\sim}{c}$ in terms of $\overrightarrow{OC}$ and $\overrightarrow{AB}$.	
7 State the conclusion.	

p. 27

CHAPTER

2 RATIONAL FUNCTIONS

p. 39

MATCHED EXAMPLE 1 Finding a locus	
Find the locus of a point equidistant from the points (2, 1) and (−2, 3).	
Steps	**Working**
1 Make a sketch. Show the point $P(x, y)$ equidistant from (2, 1) and (−2, 3).	
2 Equate the distance formulas.	
3 Square both sides.	
4 Simplify.	
5 State the answer.	

MATCHED EXAMPLE 2 Finding an ellipse from given information

p. 42

2

An ellipse has its major axis parallel to the y-axis and is of length 6. The equation of the line segment from the centre to the point (3, 2) on the ellipse is given by $y = 2$ and has length 2.

Find possible equations for the ellipse.

Steps	Working
1 Use the length of the major axis to find a parameter.	
2 Write the equation of the straight line.	
3 Substitute the centre in the equation.	
4 Write the length of the line.	
6 Simplify, substitute and solve for h.	
7 Write the possible centres.	
8 Write the general formula, substitute known values and solve for b for both cases.	
9 Write the possible equations.	

p. 43

MATCHED EXAMPLE 3 Changing from Cartesian to parametric form

Express the ellipse $\frac{(y+2)^2}{36}+\frac{(x-1)^2}{25}=1$ in parametric form.

Steps	Working
1 Use $\sin^2(t)+\cos^2(t)=1$.	
2 Satisfy the identity.	
3 Use the roots. Specify the values of t to complete all the points of the ellipse.	
4 Use the positive roots.	
5 Simplify to get the parametric form.	

MATCHED EXAMPLE 4 Changing from parametric to Cartesian form

p. 44

Express the parametric equations $\begin{cases} y=-1+4\sec(t) \\ x=2+7\tan(t) \end{cases}$, $0 \le t \le 2\pi$, $t \ne \dfrac{\pi}{2}, \dfrac{3\pi}{2}$ in Cartesian form.

Steps	Working
1 Isolate sec (t) and tan (t).	
2 Write the identity.	
3 Substitute the expressions.	
4 Simplify.	

2

p. 45

MATCHED EXAMPLE 5	Changing from polar to Cartesian form
Write the equation of $r = \dfrac{10}{3-2\cos(\theta)}, 0 \le \theta < 2\pi$ in Cartesian form.	
Steps	**Working**
1 Divide by 3 to get e and so the type of conic.	
2 Find the vertices.	
3 Change to Cartesian coordinates.	
4 Find a and b.	
5 Find the centre.	
6 Write and simplify the equation.	

MATCHED EXAMPLE 6	Changing from Cartesian to polar form
Write the equation $\frac{(y-4)^2}{9}-\frac{x^2}{7}=1$ in polar form.	
Steps	**Working**
1 State the type of conic.	
2 Use a and b to find e.	
3 Check that the centre is at a focus.	
4 Find d and de.	
5 Write and simplify the equation.	
6 Choose the correct sign.	
7 Write the answer.	

p. 46

2

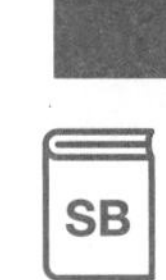

Using CAS 1: Graphing conics p. 46

p. 50

MATCHED EXAMPLE 7	Points of discontinuity
For each function, find any points of discontinuity. **a** $f(x)=\frac{x^2+5x+4}{x^2+x+2}$ **b** $g(x)=\frac{x+1}{x(x-1)}$ **c** $h(x)=\frac{x^2-2x+1}{x^2-5x+6}$	
Steps	**Working**
a **1** Factorise where possible. **2** Check the discriminant of the denominator. **3** State the conclusion.	
b **1** Find when the denominator is zero. **2** State the points of discontinuity.	
c **1** Factorise where possible. **2** State the points of discontinuity.	

MATCHED EXAMPLE 8 Roots of rational functions

SB p. 51

2

Find the zeros of each function.

a $f(x)=\dfrac{8x^3-24x^2-32x+96}{x^2+3x-10}$ **b** $g(x)=\dfrac{3x^2-7x-6}{x^2+x-2}$

Steps	Working
a **1** Factorise where possible.	
2 Find the roots. $x = 2$ is not a root because $f(x)$ is undefined at $x = 2$.	
b **1** Factorise where possible.	
2 Find the roots of the numerator.	

p. 52

MATCHED EXAMPLE 9	Quotient and remainder where deg (p) = deg (q)
Find the quotient and remainder of $\dfrac{12x^2-6x+36}{3x^2+2x+6}$.	
Steps	**Working**
1 Divide the highest powers: $12x^2 \div 3x^2 = 4$.	
2 Multiply the quotient (4) by the divisor. Subtract to get the remainder. $3x^2$ doesn't go into $-14x$, so stop dividing.	
3 Write the answer.	

MATCHED EXAMPLE 10	Quotient and remainder where deg (p) > deg (q)
Find the quotient and remainder of $\frac{6x^3-13x^2+17x-17}{2x^2-x+3}$.	
Steps	**Working**
1 Set out as a long division, with all powers shown.	
2 Write the answer.	

SB p. 52

SB Using CAS 2: Quotients of rational functions p. 52

p. 53

MATCHED EXAMPLE 11	Expression as partial fractions

Express each expression as partial fractions.

a $\dfrac{5x-4}{x^2-x-2}$ **b** $\dfrac{4x^2+5x+8}{(x^2+5)(x+2)}$

Steps	**Working**
a **1** Factorise the denominator.	
2 Decompose the denominator.	
3 Express with a common denominator.	
4 Simplify and equate the denominators.	
Method 1	**Method 2**
5 Put $x = 2$. $10 - 4 = 2A - 2A + 2B + B$	
6 Simplify. $3B = 6$ $B = 2$	
7 Put $x = -1$. $-5 - 4 = -A - 2A - B + B$	
8 Simplify. $-3A = -9$ $A = 3$	
9 Write the result.	
b **1** Check $x^2 + 5$.	
2 Write the conclusion.	
3 Decompose the denominator.	
4 Express with a common denominator.	
5 Equate numerators.	
6 Put $x = -2$.	
7 Solve for A.	
8 Simplify the RHS denominator.	
9 Collect terms.	

10 Equate coefficients and solve.	
11 Write the result.	

2

Using CAS 3: Partial fractions p. 55

p. 58

MATCHED EXAMPLE 12 Graphing reciprocal polynomial functions

Sketch the graph of $f(x)=\frac{1}{x^2+2x-8}$.

Steps	Working
1 Factorise $p(x)=x^2+2x-8$.	
2 State the important points of $p(x)$.	
3 State the signs of $p(x)$.	
4 Use $p(x)$ to get characteristics of $f(x)$.	
5 Sketch the graph, labelling all asymptotes and intercepts with their coordinates. Label the maximum.	

MATCHED EXAMPLE 13	Graphing a multiple of a reciprocal polynomial function

Sketch the graph of $f(x)=\frac{-18}{x^2+3x+6}$.

Steps	**Working**
1 Consider $p(x) = x^2 + 3x + 6$.	
2 Complete the square for $p(x)$.	
3 State the behaviour as $x \to \pm\infty$.	
4 Find the y-intercept.	
5 Sketch the graph, labelling all asymptotes and intercepts with their coordinates. Label the minimum.	

SB p. 59

p. 59

MATCHED EXAMPLE 14 Graphing a rational function with a linear numerator

Sketch the graph of $f(x)=\frac{4x+5}{x^2-1}$.

Steps	Working
1 Consider $p(x)=x^2-1$.	
2 Find any zeros.	
3 Find the y-intercept.	
4 Differentiate.	
5 Find any stationary points.	
6 State the behaviour as $x \to \pm\infty$.	
7 State the signs of $f(x)$ between the asymptotes and zeros.	
8 Sketch the graph, labelling all asymptotes and intercepts with their coordinates. Label the maximum.	

MATCHED EXAMPLE 15 Graphing a rational function with a quadratic numerator

SB

p. 60

Sketch the graph of $f(x)=\dfrac{x^2-16}{(x+2)(x+3)(x-1)}$.

Steps	Working
1 Factorise.	
2 State the zeros and vertical asymptotes.	
3 State the signs of the function.	
4 State the behaviour as $x \to \pm\infty$.	
5 Find the y-intercept.	
6 Sketch the graph, labelling all asymptotes and intercepts with their coordinates.	

2

Using CAS 4: Rational functions p. 61

p. 65

MATCHED EXAMPLE 16 Quotient function graph with a horizontal asymptote

Sketch the graph of $f(x)=\dfrac{4x^2-9}{(x+2)(x-2)}$.

Steps	Working
1 Factorise.	
2 State the zeros and vertical asymptotes.	
3 Find the y-intercept.	
4 Find the horizontal asymptote by dividing the leading coefficients.	
5 Differentiate and simplify to find stationary points.	
6 Find any points where $f'(x)=0$.	
7 Find the nature of the stationary points.	
8 State the nature of the stationary point.	
9 State the sign of $f(x)$ in each region of the graph.	
10 State the behaviour as $x \to \pm\infty$ [e.g., try $f(100), f(-100)$].	
11 Sketch the graph, marking all the important points.	

MATCHED EXAMPLE 17	Quotient function graph with a sloping asymptote

Sketch the graph of $f(x)=\dfrac{(x+2)(1-x)(x+3)}{x^2-3x+2}$.

Steps	Working
1 Factorise.	
2 State the zeros and vertical asymptotes.	
3 State the signs of the function.	
4 Find the sloping asymptote.	
5 State the behaviour as $x \to \pm\infty$.	
6 Find the y-intercept.	
7 Sketch the graph.	

SB p. 67

p. 68

MATCHED EXAMPLE 18	Quotient function graph with a curved asymptote
Sketch the graph of $f(x)=\dfrac{(x+3)(x-6)(x-1)}{x-4}$.	
Steps	**Working**
1 State the zeros and vertical asymptotes.	
2 State the signs of the function.	
3 Find the curved asymptote.	
4 State the behaviour as $x \to \pm\infty$.	
5 Find the y-intercept.	
6 Determine the characteristics of the curved asymptote.	
7 Sketch the graph.	

Using CAS 5: Quotient functions p. 69

MATCHED EXAMPLE 19	**Absolute value equations and inequalities**
Solve **a** $\|4-6x\|=16$ **b** $\|2x-6\|\geq 4$ **c** $\|x-1\|<5$	
Steps	**Working**
a **1** If $\|a\| = 16$, then $a = 16$ or -16. **2** Solve each equation. **3** Write the solutions.	
b **1** If $\|a\| \geq 4$, then $a \geq 4$ or $a \leq -4$. **2** Solve each inequality. **3** Write the solutions.	
c **1** If $\|a\| < 5$, then $-5 < a < 5$. **2** Solve each inequality. **3** Write the solution.	

SB p. 73

SB Using CAS 6: Absolute value equations and inequalities p. 74

p. 74

MATCHED EXAMPLE 20	Absolute value graphs
Sketch the graph of $f(x) = \|2x + 1\| - 3$.	
Steps	**Working**
1 State the transformations.	
2 Sketch the graph and include a few points for reference.	

MATCHED EXAMPLE 21 Graphing reciprocal circular functions

Sketch the graph of $y = 4\sec\left(2x + \frac{\pi}{2}\right) + 1$ from $x = 0$ to $x = 2\pi$.

SB

p. 78

Steps	Working
1 Express as $af[n(x+b)]+c$.	
2 State the transformations.	
3 State the asymptotes and important points of sec (x) (from zeros, maxima and minima of sin (x)).	
4 Transform the x values.	
5 Transform the y values.	
6 State the change of period.	
7 State the asymptotes and important points of the new function.	
8 Sketch the graph over the required domain, labelling important points.	

Using CAS 7: Reciprocal circular functions p. 79

p. 80

MATCHED EXAMPLE 22 Equations involving reciprocal circular functions

Solve $\sec(2x)+\sqrt{2}=0$ for all real values of x.

Steps	Working
1 Change sec to cos.	
2 Subtract $\sqrt{2}$ from both sides.	
3 Multiply $\cos(2x)$ on both sides.	
4 Divide by $-\sqrt{2}$.	
5 Solve for $\cos(2x)$.	
6 Solve for $2x$.	
7 Solve for x.	

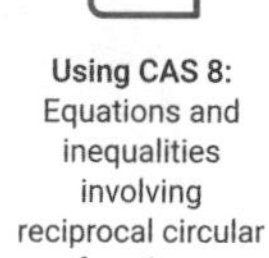

Using CAS 8: Equations and inequalities involving reciprocal circular functions p. 80

MATCHED EXAMPLE 23 Compound angles and exact values

SB p. 85

Show that $\cos\left(\frac{7\pi}{12}\right) = \frac{\sqrt{2}-\sqrt{6}}{4}$, and hence that $\sec\left(\frac{7\pi}{12}\right) = -\sqrt{2}-\sqrt{6}$.

Steps	Working
1 Choose an appropriate sum or difference identity.	
2 Expand using $\cos(x+y)$.	
3 Replace the sines and cosines with their exact values and simplify.	
4 Invert to get $\sec\left(\frac{7\pi}{12}\right)$	
5 Rationalise the denominator.	
6 Cancel to get the answer.	

p. 86

MATCHED EXAMPLE 24	Double angle application to trigonometric equation
Solve $\sqrt{3}\,\text{cosec}(2x) = 2 + \sqrt{3} - 2\sin(2x)$ for all real values of x.	
Steps	**Working**
1 Change cosec to sine.	
2 Simplify.	
3 Write in standard quadratic form.	
4 Factorise the quadratic.	
5 Solve for $\sin(2x)$.	
6 Solve for $2x$.	
7 Solve for x.	

MATCHED EXAMPLE 25 Identities and exact values

SB p. 86

2

Given $\sin(x) = \frac{\sqrt{3}}{2}$ and $\frac{\pi}{2} < x < \frac{3\pi}{2}$, find the value of $\cot(x)$.

Steps	Working
1 Find cosec (x).	
2 Now, find cot (x).	
3 Find the quadrant.	
4 Exclude the incorrect answer.	

p. 87

MATCHED EXAMPLE 26 Identities and equations

Solve $x^2 - x - 6\sin^2(x) = 6\cos^2(x)$.

Steps	Working
1 Rearrange the equation to get the identity $\sin^2(x) + \cos^2(x) = 1$	
2 Substitute in the equation.	
3 Factorise.	
4 Use the null factor law.	
5 Write the solutions.	

MATCHED EXAMPLE 27	Exact value of an inverse circular function
Find the value of **a** $\sin^{-1}(-1)$ **b** $\text{arcsec}\left(\sqrt{2}\right)$.	
Steps	**Working**
a **1** What value of $\sin(y) = 1$?	
2 For what value does $\sin(y) = -1$ with y in the first three quadrants?	
3 Write the answer.	
b **1** Write in terms of $\arccos(x)$.	
2 What angle has $\cos(x) = \frac{1}{\sqrt{2}}$?	

p. 89

Using CAS 9: Inverse circular functions p. 90

p. 91

MATCHED EXAMPLE 28 Domain and range of inverse circular functions

What are the maximal domain and range of $2\cos^{-1}\left(\frac{x}{4}+2\right)-\frac{\pi}{2}$?

Steps	Working
1 Use the domain of $\cos^{-1}(A)$.	
2 Solve each side.	
3 Combine the solutions.	
4 Use the range of $\cos^{-1}(A)$.	
5 Multiply by 2.	
6 Subtract $\frac{\pi}{2}$.	
7 Write the answer.	

MATCHED EXAMPLE 29	Transformation of inverse circular functions
Sketch the graph of $y=3\arccos(x-4)+2$.	
Steps	**Working**
1 State the transformations.	
2 Do the horizontal translation.	
3 Do the vertical dilation.	
4 Do the vertical translation.	

SB

p. 92

p. 93

MATCHED EXAMPLE 30	Transformation of arctan function
Sketch the graph of $y = 3\tan^{-1}(-3x + 2) + 3$.	
Steps	**Working**
1 Write in the standard form.	
2 State the transformations.	
3 Transform the important points and asymptotes. Only the vertical changes are needed on the asymptotes.	
4 Sketch the graph.	

Using CAS 10: Graphing inverse circular functions p. 94

p. 95

MATCHED EXAMPLE 31 Implied domains and ranges of inverse circular functions

State the implied domain and range of each function.

a $f(x) = a + d\arccos^2(bx - c)$, where $a, b, c, d \in R$ and $b > 0$

b $y = \sqrt{a\arctan(cx-d)} + b$, where $a, b, c, d \in R$ and $a, c > 0$

Steps	Working
a 1 Use the domain of arccos (x).	
2 Solve for x.	
3 Use the range of arccos (x).	
4 Work towards the expression.	
5 Write the answer.	
b 1 Use the domain of arctan (x).	
2 The square root of $\arctan(cx-d)$ is part of the expression.	
3 Solve for x.	
4 Use the (restricted) range of arctan $(cx - d)$.	
5 Work towards expression.	
6 Write the answer.	

LOGIC AND PROOF

p. 114

MATCHED EXAMPLE 1	Identifying logical terms	
a In the conversation, decide if each sentence is a statement.		
b Describe the statement that is not atomic.		
c Describe each inference.		
d Explain if there is a logical argument.		
Chen: How was the chess tournament, Lucy?		1
Lucy: I won.		2
Chen: That is because your strategic thinking skills are good.		3
Lucy: I won the tournament because I played chess every day the last year.		4
Chen: Playing chess every day helps to develop strategic thinking skills.		5

Steps	**Working**
a A statement is true or false.	
b Look for a sentence that can be written as two or more sentences.	
c Identify a premise from which a conclusion is made.	
d Decide if the premises and inferences are correct and that a general conclusion is made.	

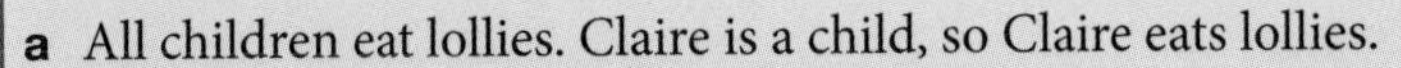

MATCHED EXAMPLE 2 Inductive and deductive reasoning

p. 115

3

Decide, with reasons, whether each conclusion is made through inductive reasoning, deductive reasoning or neither.

a All children eat lollies. Claire is a child, so Claire eats lollies.

b The farmer's market was open every Friday in the last month. It is Friday today, so the farmer's market should be open.

c Every time you eat peanuts, you start to cough. You are coughing, so you are eating nuts.

Steps	Working
a **1** Write the information consisting of statements and a conclusion.	
2 Decide if the information fits the definition of inductive or deductive reasoning.	
3 State the type of reasoning.	
b **1** Write the information consisting of statements and a conclusion.	
2 Decide if the information fits the definition of inductive or deductive reasoning.	
3 State the type of reasoning.	
c **1** Write the information consisting of statements and a conclusion.	
2 Decide if the information fits the definition of inductive or deductive reasoning.	
3 State the type of reasoning.	

p. 116

MATCHED EXAMPLE 3 Conjectures and counterexamples

For each conjecture **i** provide an example when the conjecture is true and **ii** give a counterexample.

a The sum of the squares of any two integers is an even number.

b For all integers a and b, $\frac{a}{b} > \frac{b}{a}$.

c For all natural numbers, the remainder when divided by 3 is zero.

Steps	Working
a **i** State a particular case when the conjecture holds. **ii** Provide a counterexample.	
b **i** Give an example when the conjecture holds. **ii** Give a counterexample.	
c **i** Provide an example that shows the conjecture to be true. **ii** Give a counterexample.	

MATCHED EXAMPLE 4 Forming a conjecture

SB p. 117

a State a conjecture using notation that describes the sequence $\frac{8}{3}, \frac{14}{3}, \frac{20}{3}, \ldots$

b Predict the next three terms of the sequence.

c Provide an informal justification of your conjecture.

Steps	Working
a **1** Look for patterns.	
2 Provide a conjecture.	
b Predict more terms.	
c Show general relationships.	

3

p. 122

MATCHED EXAMPLE 5 Logical connectives

Let A, B, C and D be the statements:

A: 'Dogs are good pets'

B: 'Dogs are trained'

C: 'I like dogs'

D: 'Dogs are loyal'.

Write each symbolic statement using correct grammar.

a $A \rightarrow B$ **b** $\neg D \rightarrow \neg A$ **c** $(D \wedge B) \rightarrow C$ **d** $\neg D \rightarrow \neg B \vee \neg A$

e $C \leftrightarrow B$

Steps	Working
a **1** Write the statement using word connectives. **2** Write the completed sentence.	
b **1** Write the statement using word connectives. **2** Write the completed sentence.	
c **1** Add word connectives to the statements. **2** Write the completed sentence.	
d **1** Add word connectives. **2** Write the completed sentence.	
e **1** Join statements with word connectives. **2** Write the completed sentence.	

MATCHED EXAMPLE 6 Statements using symbolic logic

p. 123

Write statements with logical connectives to describe each of the following.

a If I knew how to cook, I would have made pasta and invited you to lunch.

Use A: know to cook. B: made pasta. C: invite you to lunch.

b If a number is real, then it is either rational or irrational.

Use A: number is real. B: number is rational. C: number is irrational.

c If Susan studies mathematics and does not miss lectures, then she does not fail and gets the scholarship.

Use A: Susan studies mathematics. B: Susan misses lectures. C: Susan fails.

D: Susan gets the scholarship.

Steps	Working
a **1** Decide on the connectives. **2** Write the information in notation form.	
b **1** Decide on the connectives. **2** Write the information in notation form.	
c **1** Decide on the connectives to use. **2** Write the sentence in logical form.	

p. 124

MATCHED EXAMPLE 7 Definitions as biconditional statements

Express each definition as a biconditional statement.

a A right angle has a measure of 90°.

b Anita goes to the library on a weekday.

Steps	Working
a **1** Find statements A and B so that $A \rightarrow B$ and $B \rightarrow A$.	
2 State $A \rightarrow B$ and $B \rightarrow A$ in words.	
3 State the biconditional statement $A \leftrightarrow B$.	
b **1** Find statements A and B so that $A \rightarrow B$ and $B \rightarrow A$.	
2 State $A \rightarrow B$ and $B \rightarrow A$ in words.	
3 State the biconditional statement $A \leftrightarrow B$.	

MATCHED EXAMPLE 8 'if and only if' involving divisibility

Show that for $n \in N$, n is divisible by both 2 and 3 if and only if n is divisible 6.

Steps	Working
1 Choose statements.	
2 Show $X \rightarrow Y$.	
3 Show $Y \rightarrow X$.	

SB

p. 124

p. 125

MATCHED EXAMPLE 9	'if and only if' involving square numbers
Show that for $n \in N$, $n^2 + 2$ odd $\leftrightarrow n$ is odd.	
Steps	**Working**
1 Choose statements.	
2 Show $X \rightarrow Y$.	
3 Show $Y \rightarrow X$.	

SB

p. 126

3

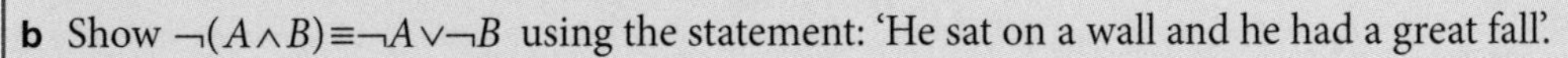

MATCHED EXAMPLE 10 Verifying De Morgan's laws

a From the set of numbers {1, 2, 3, 4, 5, 6, 7, 8, 9, 10, 11, 12, 13, 14, 15}, A chooses all the numbers that are divisible by 3, and B chooses all the numbers that are divisible by 2.
Show $\neg(A \vee B) \equiv \neg A \wedge \neg B$ is true.

b Show $\neg(A \wedge B) \equiv \neg A \vee \neg B$ using the statement: 'He sat on a wall and he had a great fall'.

c Show that $\neg(A \vee B \vee C) \equiv \neg A \wedge \neg B \wedge \neg C$.

Steps	Working
a **1** Write the statements to be used.	
2 Find the set of numbers that satisfy $\neg(A \vee B)$.	
3 Find the set of numbers that satisfy $\neg A \wedge \neg B$.	
b **1** Write the statements to use.	
2 Express $\neg(A \wedge B)$ with words.	
3 Express $\neg(A \wedge B)$ with correct grammar.	
4 Express $\neg A \vee \neg B$ in word form.	
5 Express $\neg A \vee \neg B$ using correct grammar.	
c **1** Write in the form $\neg(X \vee Y)$	
2 Apply De Morgan's law twice.	

p. 129

MATCHED EXAMPLE 11 Direct proof

a Prove that if an integer n is even, then n^2+2n+4 is divisible by 4.

b Given that a^n-b^n is divisible by $a-b$ for integer n, prove that for any integer n, 8^n-1 is divisible by 7.

Steps	Working
a **1** State what is to be proved.	
2 Write the first statement using known information.	
3 Establish the conclusion.	
b **1** Write the statements to use.	
2 Write the first statement using known information.	
3 Establish the conclusion.	

MATCHED EXAMPLE 12 Interpreting expressions containing quantifiers

p. 130

Write each expression as a statement. If no domain is given, assume it is R.

a $\forall x \in Z, \exists y \in Z\left(y = x^2 + 1\right)$

b $\forall p, q \neq 0 \in Z, \exists y \in R\left(y = \dfrac{p}{q}\right)$

Steps	Working
a **1** State the meaning of each quantifier and the predicate.	
2 Form one statement.	
b **1** State the meaning of each quantifier and the predicate.	
2 Form one statement.	

p. 131

MATCHED EXAMPLE 13 Writing statements using quantifiers

Write each expression using appropriate quantifiers. If no domain is given, it can be assumed that it is R and need not be included as part of the answer.

a $y=\frac{1}{x}$ **b** $a=\cos(b)$

Steps	Working
a **1** Determine the domain, quantifiers and predicate.	
2 Form one statement.	
b **1** Determine the domain, quantifiers and predicate.	
2 Form one statement.	

MATCHED EXAMPLE 14	Decide if statements involving quantifiers are true
Decide if each of the following statements is true or false. Assume x and y are natural numbers.	
a $\forall x\,(x>8 \rightarrow \exists y\,(x+y=10))$	**b** $\forall x(x<4 \rightarrow \exists y(y=x^2))$
Steps	**Working**
a **1** Work from left to right and test each given condition.	
2 Decide if the statement is true.	
b **1** Write the statement as a sentence.	
2 Decide if the statement is true.	

p. 131

p. 132

MATCHED EXAMPLE 15 Proof using quantifiers 1

Prove $\forall x, y \in Q((x \cdot y) \in Q)$

Steps	Working
1 Translate the statement to omit quantifiers.	
2 Write as a conditional statement, $P \rightarrow Q$.	
3 Use P to deduce Q.	
4 State the proof.	

MATCHED EXAMPLE 16 Proof using quantifiers 2

p. 132

Prove that each of the following statements is false.

a There is a positive integer, n, such that $n^2 + 7n + 12$ is prime.

b Every integer is even.

Steps	Working
a **1** Write the statement using quantifiers.	
2 Write the negation.	
3 Show that the statement is true.	
b **1** Write the statement using quantifiers.	
2 Write the negation.	
3 Show the statement to be true.	

p. 135

MATCHED EXAMPLE 17	Proof by contrapositive
Prove that $\forall n \in N$, if $(-1)^n = 1$, then n is even.	
Steps	**Working**
1 Express the statement $P \to Q$ in the form $\neg Q \to \neg P$.	
2 Prove $\neg Q \to \neg P$.	

MATCHED EXAMPLE 18 Contrapositive to prove divisibility

p. 135

Prove that if $n^2 - 1$ is not divisible by 8, then n is not an odd positive integer.

Steps	Working
1 Express the statement $P \to Q$ in the form $\neg Q \to \neg P$.	
2 State $\neg Q \to \neg P$	
3 Prove $\neg Q \to \neg P$	

p. 135

MATCHED EXAMPLE 19 Proof by contradiction involving surds

Use proof by contradiction to prove that $\sqrt{5}$ is irrational.

Steps	Working
1 Assume the statement is false and write it as an equation.	
2 Rearrange the equation and form conclusions about the types of variables used.	
3 Write the equation in terms of the types of variables and rearrange.	
4 State the contradiction.	
5 State the conclusion.	

9780170464079

MATCHED EXAMPLE 20	Proof by contradiction
Show that for integer a if a^2 is even then a is even.	
Steps	**Working**
1 Assume the statement is false.	
2 Rewrite the expression according to the types of numbers used.	
3 Expand and rearrange the expression and form a conclusion.	
4 State what must be true to avoid a contradiction.	
5 Explain the contradiction, testing each case.	
6 State the conclusion.	

p. 136

3

p. 139

MATCHED EXAMPLE 21 Proof by induction for the sum of terms of an arithmetic sequence

Prove by induction that $3+7+11+...+(4n-1)=n(2n+1)$, $n \in N$.

Steps	Working
1 Prove the base step.	
2 State the hypothesis and the required expression.	
3 Write the function in the required form.	
4 State the conclusion.	

MATCHED EXAMPLE 22 Proof by induction of the sum of terms of square numbers

SB p. 139

Prove by induction that $1+4+9+16+..+n^2=\frac{n}{6}(n+1)(2n+1),\ n\in N$.

Steps	Working
1 Prove the base step.	
2 State the hypothesis and the required expression.	
3 Write the function in the required form.	
4 State the conclusion.	

p. 140

MATCHED EXAMPLE 23	**Divisibility proof using induction**
Prove by induction that for positive integer n, $3^{2n}-1$ is divisible by 8.	
Steps	**Working**
1 Prove the base step.	
2 State the hypothesis and the required expression.	
3 Write the function as the multiple of 8.	
4 State the conclusion.	

MATCHED EXAMPLE 24 Induction with trigonometry

SB p. 141

Given $n \in N$ and $\sin\left(\frac{x}{2}\right) \neq 0$, show $\cos\left(\frac{x}{2}\right) + \cos\left(\frac{3x}{2}\right) + \cos\left(\frac{5x}{2}\right) + .. + \cos\left(\frac{2n-1}{2}\right)x = \frac{\sin(nx)}{2\sin\left(\frac{x}{2}\right)}$.

Steps	Working
1 Prove the base step.	
2 State the hypothesis and the required expression.	
3 Write the function in the required form.	

4 State the conclusion.

COMPLEX NUMBERS

MATCHED EXAMPLE 1	Complex numbers in the complex plane
a Show $u = -3 + 5i$ and $v = 3 + 5i$, as points on an Argand diagram.	
b What is the relationship between u and v?	
c Show $w = 3 - 2i$ and $z = 2 + 3i$, as vectors on an Argand diagram.	
d What is the relationship between w and z?	

p. 152

Steps	Working
a **1** Write the complex numbers as points. **2** Plot and label the points on the plane.	
b State the relationship.	
c **1** Write the complex numbers as points.	

2 Draw as position vectors.

d State the relationship.

MATCHED EXAMPLE 2 Complex number operations

Simplify each expression.

a $(1-3i)+(2-5i)$ **b** $(-3+5i)-(1-4i)$ **c** $(-3-i)\times(2+3i)$ **d** $(3-2i)\div(5+i)$

SB p. 152

Steps	Working
a **1** Separate the real and complex parts.	
2 Add the parts.	
b **1** Separate the real and complex parts.	
2 Complete the subtractions.	
c **1** Expand the brackets.	
2 Multiply out.	
3 Use $i^2=-1$.	
4 Express in standard form.	
d **1** Write in fraction form.	
2 Multiply by $\frac{\overline{w}}{\overline{w}}(=1)$ and simplify the expression.	
3 Express as $x+yi$. 	

Using CAS 1: Complex number operations p. 153

p. 156

MATCHED EXAMPLE 3 Polar form

$u = 4\operatorname{cis}\left(\frac{\pi}{4}\right)$, $v = 2\sqrt{3} + 2i$, $w = 2\operatorname{cis}\left(-\frac{\pi}{6}\right)$ and $z = -2i - 2$

a Express v and z in polar form.

b Express u and w in Cartesian form.

c Find uv and wv.

d Find $\frac{v}{u}$ and $\frac{z}{u}$.

Steps	Working
a **1** Find the modulus and principal argument of v.	
2 Find the modulus and principal argument of z.	
3 Write the polar forms.	
b **1** Write in complete form.	
2 Substitute the values of the trig ratios and multiply out.	
c **1** Write uv in polar form, multiply the moduli and add the arguments.	
2 Write wv in polar form, multiply the moduli and add the arguments.	
d **1** Write $\frac{v}{u}$ in polar form, divide the moduli and subtract the arguments.	
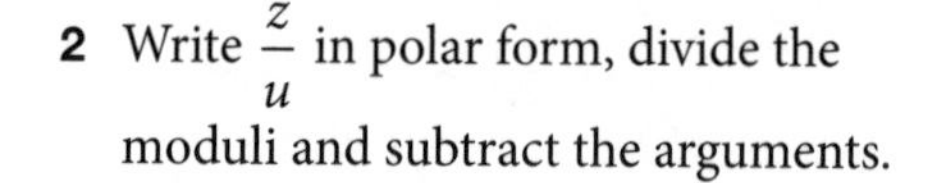 **2** Write $\frac{z}{u}$ in polar form, divide the moduli and subtract the arguments.	

Using CAS 2: Polar and Cartesian conversions p. 158

MATCHED EXAMPLE 4	**Straight lines in the complex plane**
a What is the equation of the line passing through $3 - 2i$ and $1 + 2i$?	
b Express the equation in Cartesian form.	

p. 159

Steps	**Working**
a **1** Use $z - u = k(v - u)$.	
2 Simplify.	
3 Write the answer.	
b **1** Write $z = x + iy$.	
2 Equate real and imaginary parts.	
3 Eliminate k.	
4 Write the answer.	

p. 159

MATCHED EXAMPLE 5 Circles in the complex plane

a Find the equation of the circle with centre $-2 + 3i$ and radius 4.

b Sketch and label the region strictly outside the circle.

Steps	Working
a **1** Use $\lvert z - u \rvert = r$.	
2 Simplify.	
b **1** The inequation will use > for strictly outside the circle.	
2 Sketch the circle with a dashed line to show the circumference is not included and shade the region outside the circle.	

MATCHED EXAMPLE 6 Ellipses in the complex plane

A region of the complex plane is given by $|z + 1 + 4i| + |z - 2 - 5i| \leq 10$.

Describe and sketch the shape.

p. 160

Steps	Working
1 Describe the shape.	
2 Use the foci and $a = 5$. Include the circumference.	

4

p. 161

MATCHED EXAMPLE 7 Semicircles in the complex plane

A region of the complex plane is given by $\frac{\pi}{2} < \text{Arg}\,(z + 2 - 3i) - \text{Arg}\,(z - 5 - i) \leq \pi$.
Describe and sketch the shape.

Steps	Working
1 Describe the shape.	
2 Sketch the shape.	

MATCHED EXAMPLE 8	**Combined regions**

A region of the complex plane is given $\{z: z - \overline{z} \le 6i\} \cap \{z: \frac{\pi}{3} \le \text{Arg}(z - 2 + 4i) < \frac{3\pi}{4}\}$.

a Describe and sketch the shape.

b Find its area.

p. 161

4

Steps	Working
a **1** Describe the shape.	
2 Consider the regions.	
3 Sketch the graph, including the line $z = 3i$ and the ray $\frac{\pi}{3} = \text{Arg}(z - 2 + 4i)$, but excluding the other ray.	
b **1** Draw a diagram.	
2 Find the height and base.	
3 Find the area.	

p. 169

MATCHED EXAMPLE 9	Power of a complex number
Given $z=\sqrt{6}-3\sqrt{2}i$, write z in polar form and hence find z^4 and express it in Cartesian form.	
Steps	**Working**
1 Find the modulus and argument.	
2 Write in polar form.	
3 Find z^4 using de Moivre's theorem.	

Using CAS 3: Powers of complex numbers p. 170

p. 172

MATCHED EXAMPLE 10	**Complex roots of unity**
Solve $z^5 = 1$ and show the solutions on an Argand diagram.	
Steps	**Working**
1 The modulus of z will be 1.	
2 Use de Moivre's theorem.	
3 Write the equation with $1 = \cos(2k\pi) + i\sin(2k\pi)$.	
4 Write the argument equation and solve for θ.	
5 Write out some solutions.	
6 Choose only the principal arguments.	
7 Write the solutions.	
8 Show the roots on an Argand diagram.	

p. 173

MATCHED EXAMPLE 11	General complex roots
Find the cube roots of $z = -32\sqrt{2} - 32i\sqrt{2}$ and show them on an Argand diagram.	
Steps	**Working**
1 Find z in polar form.	
2 Let the roots be w and write $w^3 = z$ in general form.	
3 Use de Moivre's theorem.	
4 Choose only the principal arguments.	
5 Show the roots on an Argand diagram.	

Using CAS 4: Roots of unity p. 174

p. 177

MATCHED EXAMPLE 12 Factorisation of quadratics with complex factors

Factorise each quadratic expression over the complex numbers.

a $z^2 - 4z + 6$ **b** $3z^2 + 4z + 2$

Steps	Working
a **1** Complete the square. This can also be done with the quadratic formula.	
2 Use i to make a difference of squares.	
3 Complete the factorisation.	
b **1** Use the quadratic formula to find the roots of $3z^2 + 4z + 2 = 0$. This can also be solved by completing the square.	
2 Use the roots for the factorisation and simplify.	

p. 178

Using CAS 5: Polynomial division remainders p. 178

MATCHED EXAMPLE 13 Finding a remainder

Find the remainder when $p(z) = z^4 + (1 - 3i)z^3 - 4z + 6 - i$ is divided by $z + 1 - 2i$.

Steps	Working
1 Apply the remainder theorem by substituting $(-1 + 2i)$ in $p(z)$.	
2 Write the answer.	

MATCHED EXAMPLE 14	**Finding factors**
Show that $z - 5 - i$ and $z - 3$ are both factors of $P(z) = z^3 - (5 + 3i)z^2 - (11 - 16i)z + 51 - 21i$ and find the remaining factor.	

SB p. 178

4

Steps	Working
1 Apply the factor theorem by substituting $(5 + i)$ in $P(z)$ and showing that it is equal to 0.	
2 Substitute 3 in $P(z)$ and show it is equal to 0.	
3 State the result.	
4 Identify the remaining factor.	
5 Write $P(z)$ as a product.	
6 Find a.	
7 Write the remaining factor.	

p. 180

MATCHED EXAMPLE 15 Factorising a complex quadratic

Factorise:

a $z^2 + 4i$ **b** $z^2 + (4 - 2i)z + 3 - 2i$

Steps	Working
a **1** Write as $z^2 = \ldots$ Write in general polar form to make it easier to find the square roots. **2** Use de Moivre's theorem to find the general roots. **3** Choose only the principal arguments. **4** Use the roots to write the answer.	
b **1** Try the factors of $3 - 2i$. **2** Use the factor theorem. **3** Write $p(z)$ as factors. **4** Expand. **5** Equate real and imaginary parts to solve for a and b. **6** Write the answer.	

MATCHED EXAMPLE 16	Factorising a complex cubic

p. 181

Factorise

a $z^3 + z^2 i - z^2 - 4z + 4 - 4i$ **b** $z^3 - 2z^2 + (10 - 3i)z - 9 + 3i$.

Steps	Working
a 1 Group terms to get a common factor.	
Factorise.	
b 1 Try some values to get $p(a) = 0$.	
2 State the factor.	
3 Write $p(z)$ as a product of factors and expand.	
4 Use the constant term to find b.	
5 Use the z^2 term to find a.	
6 Check using the z term.	
7 Write the partial result.	
8 Try some values to get $q(a) = 0$.	
9 State the factor.	
10 Write $q(z)$ as a product of factors and expand.	
11 Use the constant term to find c.	
12 Write the answer.	

p. 184

MATCHED EXAMPLE 17	Real quadratic equations with complex solutions		
Solve each quadratic equation.			
a $z^2 + 81 = 0$	**b** $z^2 - 4z + 5 = 0$	**c** $2z^2 + 7z + 14 = 0$	

Steps	**Working**
a 1 Rearrange.	
2 Separate -1 and simplify to get i.	
3 Write the answer.	
b 1 Solve by completing the square. This can also be solved by using the quadratic formula.	
2 Write $\sqrt{-1}$ as i and solve.	
3 Write the answer.	
c 1 Solve using the quadratic formula. This can also be solved by completing the square.	
2 Simplify the surd.	
3 Write $\sqrt{-7}$ as $\sqrt{7}\,i$ and simplify.	
4 Write the answer.	

MATCHED EXAMPLE 18	Finding a real quadratic equation from a complex root

One of the roots of a quadratic equation with real coefficients is $z = \frac{1}{2} - \sqrt{3}i$. Find a possible equation.

SB

p. 185

4

Steps	Working
1 Write the other root.	
2 Write a quadratic equation with these roots.	
3 Expand and simplify.	
4 Find $z + \overline{z}$.	
5 Find $z\overline{z}$.	
6 Substitute $z + \overline{z}$ and $z\overline{z}$ into the equation and simplify.	

p. 185

MATCHED EXAMPLE 19 Complex quadratic equations

Solve the equations:

a $2iz^2 + 14 = 0$ **b** $2iz^2 + 5z + 3i = 0$ **c** $z^2 + 2\sqrt{3}iz - 3\sqrt{3}i = 0$

Steps	Working
a **1** The coefficient of z^2 includes i, so multiply by i.	
2 Write as $z^2 = \ldots$	
3 Write in general polar form.	
4 Use de Moivre's theorem.	
5 Use only the principal roots.	
6 Change to Cartesian form and simplify.	
b **1** The coefficient of z^2 includes i, so multiply by i.	
2 Check the discriminant.	
3 The discriminant is real, so use the quadratic formula. Don't re-calculate the discriminant, just use $\sqrt{-49} = 7i$.	
4 Simplify.	
c **1** Check the discriminant.	
2 The discriminant is complex, so complete the square.	
3 Add the square of half the z term to each side.	
4 Write the perfect square.	
5 Write the RHS in general polar form.	

6 Use de Moivre's theorem.

7 Use only the principal values.

8 Simplify.

4

p. 187

MATCHED EXAMPLE 20	Finding a conjugate root
One root of a polynomial equation with real coefficients is $7 + 5i$. State another root of the polynomial.	
Steps	**Working**
Give the conjugate.	

MATCHED EXAMPLE 21	Solving a cubic equation with real coefficients
$p(z) = 5z^3 - 13z^2 + 56z - 30$ and $p(1 - 3i) = 0$. Solve $p(z) = 0$.	
Steps	**Working**
1 Use the factor theorem.	
2 Use the conjugate root theorem.	
3 Write $p(z)$ as a product.	
4 Solve the equation.	

SB

p. 187

4

p. 188

MATCHED EXAMPLE 22 Using the factor theorem to solve a real polynomial equation

Solve $2z^4 - 4z^3 + 9z = 7z^2 + 18$.

Steps	Working
1 Write the equation as $p(z) = 0$.	
2 Try factors of the constant -18.	
3 Write $p(z)$ as a product.	
4 Solve $p(z) = 0$, using the quadratic formula for the last factor.	
5 Write the solutions.	

MATCHED EXAMPLE 23	Solving a complex cubic equation

Solve $z^3 - 5iz^2 + (10i - 7)z - 6 + 15i = 0$.

Steps	Working
1 Write the polynomial function.	
2 Try factors of the constant $-6 + 15i$.	
3 Write $p(z)$ as a product.	
4 Solve $p(z) = 0$.	

SB p. 188

p. 189

MATCHED EXAMPLE 24	Solving a complex quartic equation
Solve $10z^4 + 49z^2 + 36 = 9iz^3 + 36iz$.	
Steps	**Working**
1 Write the polynomial function.	
2 Try factors of the constant 36.	
3 Write $p(z)$ as a product.	
4 Solve the quadratic equation.	
5 Write all the solutions of $p(z) = 0$.	

Using CAS 6: Solving polynomial equations p. 189

DIFFERENTIATION

MATCHED EXAMPLE 1 The product rule

Use the product rule to find $\frac{dy}{dx}$ for the function $y=(5x^2-12x)x^2$. Then, verify your answer by expanding the function first and then differentiating.

SB
p. 207

Steps	Working
1 Use the product rule $\frac{d}{dx}(uv)=uv'+vu'$.	
2 Simplify the answer.	
TI-Nspire	**ClassPad**
3 Verify by expanding $y=(5x^2-12x)x^2$ and then differentiating.	

Using CAS 1: Differentiation p. 205

p. 208

MATCHED EXAMPLE 2 The quotient rule

Find $\frac{dy}{dx}$ for the function $y=\frac{3x+2}{x^3+2x^2}$.

Steps | **Working**

1 Use the quotient rule $\frac{d}{dx}\left(\frac{u}{v}\right)=\frac{vu'-uv'}{v^2}$.

2 Simplify the answer.

TI-Nspire

ClassPad

MATCHED EXAMPLE 3	The chain rule
Find $\frac{dy}{dx}$ for the function $y=(4x^4+5x)^5$.	

SB

p. 208

5

Steps	Working
1 Use the chain rule $\frac{dy}{dx}=\frac{dy}{du}\times\frac{du}{dx}$.	
2 Replace $u=4x^4+5x$ and simplify the answer.	
TI-Nspire	ClassPad

SB

p. 210

MATCHED EXAMPLE 4 Gradient at a point

Find the gradient of the curve at $x = -1$ for the function $y = \frac{x^2}{2x+8}$.

Steps	**Working**
1 Use the quotient rule. $\frac{d}{dx}\left(\frac{u}{v}\right) = \frac{vu' - uv'}{v^2}$	
2 Simplify the answer.	
3 Substitute $x = -1$ into $\frac{dy}{dx}$.	

SB

Using CAS 2: Differentiation applications p. 210

MATCHED EXAMPLE 5 The chain rule with a circular function

p. 213

Find $\frac{dy}{dx}$ for the function $y = 2\cos^3(2x)$.

Steps	Working
1 Rewrite $y = 2\cos^3(2x)$ in the chain rule form.	
2 Use the chain rule $\frac{dy}{dx} = \frac{dy}{du} \times \frac{du}{dx}$ and the rule $y = \cos(kx) \Rightarrow \frac{dy}{dx} = -k\sin(kx)$.	
3 Substitute $u = \cos(2x)$.	
4 Simplify the answer using a double-angle formula.	

5

p. 214

MATCHED EXAMPLE 6	The product rule with a circular function
Find the gradient of the function $y=x^2\sin^2(3x)$ at $x=\pi$.	
Steps	**Working**
1 $y=x^2\sin^2(3x)$ in the product rule form.	
2 Use the product rule $\frac{dy}{dx}=u\frac{dv}{dx}+v\frac{du}{dx}$.	
3 Substitute $x=\pi$.	

SB

p. 214

5

MATCHED EXAMPLE 7 The quotient rule with a circular function

Find the gradient of the curve $f(x)=\dfrac{\sin^2(x)}{\cos(x)}$ at $x=\dfrac{\pi}{6}$.

Steps	Working
1 Use the quotient rule to find $f'(x)$.	
2 Substitute $x=\dfrac{\pi}{6}$	
3 Answer the question.	

p. 218

MATCHED EXAMPLE 8 Differentiating inverse circular functions

Find $\frac{dy}{dx}$ for the function $y = \cos^{-1}\left(\frac{x}{2}\right)$, giving the domain for your answer.

Steps	Working
1 Identify the value of a in $y = \cos^{-1}\left(\frac{x}{2}\right)$.	
2 Use the rule $\frac{d}{dx}\left(\cos^{-1}\left(\frac{x}{a}\right)\right) = -\frac{1}{\sqrt{a^2 - x^2}}$.	
3 Simplify your answer if necessary.	
4 Give the domain.	

MATCHED EXAMPLE 9 Differentiating inverse circular functions at a point

p. 219

Find the gradient of the function $y = \arcsin(x^2 - 1)$ at $x = \frac{1}{2}$.

Steps	Working
1 Put $y = \arcsin(x^2 - 1)$ in the chain rule form.	
2 Use the chain rule $\frac{dy}{dx} = \frac{dy}{du} \times \frac{du}{dx}$ and the rule $\arcsin(u) = \frac{dy}{du} = \frac{1}{\sqrt{1-u^2}}$.	
3 Substitute $u = x^2 - 1$.	
4 Substitute $x = \frac{1}{2}$.	

p. 222

MATCHED EXAMPLE 10	Differentiating exponential functions
Find $\frac{dy}{dx}$ for the function $y=e^{4x}x^3$ at $x=2$.	
Steps	**Working**
1 Use the product rule $\frac{dy}{dx}=u\frac{dv}{dx}+v\frac{du}{dx}$.	
2 Simplify the answer.	
3 Substitute $x=2$.	

MATCHED EXAMPLE 11	Differentiating logarithmic functions
Find $\frac{dy}{dx}$ for $y = 3x^2\log_e\left(\frac{1}{6x}\right)$, where $x \neq 0$.	
Steps	**Working**
1 Use the product rule $\frac{d}{dx}(uv) = uv' + vu'$.	
2 Simplify the answer.	

p. 222

5

Using CAS 3: Differentiation: a mixture of functions p. 224

Using CAS 4: Finding the second derivative p. 226

p. 227

MATCHED EXAMPLE 12 Second derivative

For $y = 6x^5 + 12\cos(x)$, find the following:

a $\frac{d^2y}{dx^2}$

b The second derivative at $x = 0$.

Steps	Working
a **1** Find $\frac{dy}{dx}$. **2** Find $\frac{d^2y}{dx^2}$	
b Substitute $x = 0$. TI-Nspire	ClassPad

MATCHED EXAMPLE 13	Points of inflection

For the graph of $f(x)=(2x+5)(x^2-8)$, find any points of inflection and state whether they are stationary or non-stationary.

SB

p. 230

5

Steps	Working
1 Find $f'(x)$ using the product rule.	
2 Solve $f'(x)=0$.	
3 Find $f''(x)$.	
4 Solve $f''(x)=0$.	
5 Substitute $x=-\frac{5}{6}$ into $f'(x)$.	
6 We have $f''\left(-\frac{5}{6}\right)=0$ and $f'\left(-\frac{5}{6}\right)\neq 0$.	

p. 231

MATCHED EXAMPLE 14 Concavity

For the graph of $f(x)=(x-2)(x^2+1)$, find the points of inflection and comment on the change in concavity at each point.

Steps	Working
1 Find $f'(x)$ using the product rule.	
2 Solve $f'(x) = 0$.	
3 Find $f''(x)$.	
4 Solve $f''(x) = 0$.	
5 Substitute $x = 1, \frac{1}{3}$ and $\frac{2}{3}$ into $f(x)$.	
6 Test the point of inflection at $x = \frac{2}{3}$.	
7 Discuss concavity.	

9780170464079

SB
p. 232

5

MATCHED EXAMPLE 15 The second derivative test

Show that the graph of $f(x) = 2x^3(x-2)$ has a stationary point at $x = \frac{3}{2}$ and determine its nature.

Steps	**Working**
1 Find $f'(x)$ using the product rule.	
2 Solve $f'(x) = 0$.	
3 Substitute $x = 0$ and $\frac{3}{2}$ into $f(x)$.	
4 Find $f''(x)$.	
5 Substitute $x = 0$ and $\frac{3}{2}$ into $f''(x)$.	
6 Use the second derivative test at $x = \frac{3}{2}$. 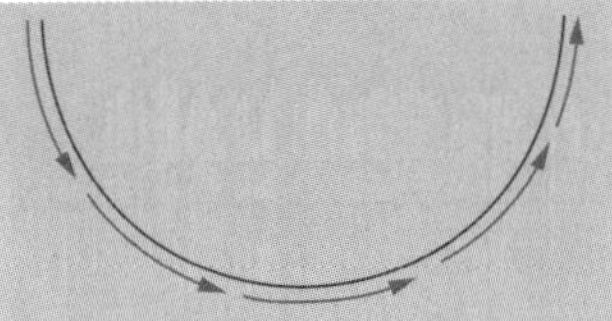	
7 State the nature of the stationary points at $x = 0$ and $x = \frac{3}{2}$.	

p. 235

MATCHED EXAMPLE 16 Related rates 1

The volume of a spherical balloon of radius r is increasing at the rate of $2\pi\ \text{m}^3$/second. At what rate will the radius increase when $r = 3$ m?

Steps	Working
1 Write the given information.	
2 Specify the derivative that needs to be determined at a given condition.	
3 Apply the chain rule using the required variables.	
4 Derive.	
5 Solve using $\frac{dV}{dt} = \frac{dV}{dr} \times \frac{dr}{dt}$.	

9780170464079

MATCHED EXAMPLE 17 Related rates 2

A right circular cone of base length 4 cm and vertical height 8 cm is being filled with water at a rate of $\frac{6}{5}$ cm^3/min. At what rate will the height of water increase when $h = 4$ cm?

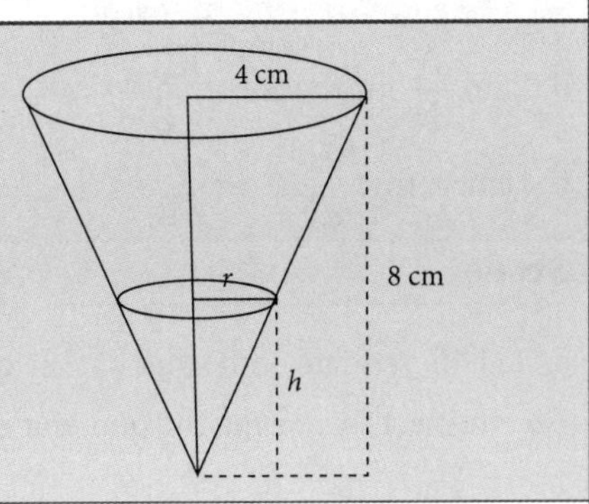

SB
p. 236

5

Steps	Working
1 Consider the given information.	
2 Specify the derivative that needs to be determined at a given condition.	
3 Apply the chain rule using the required variables.	
4 Differentiate using $\frac{5}{r} = \frac{10}{h} \therefore r = \frac{h}{2}$.	
5 Solve using $\frac{dV}{dt} = \frac{dV}{dh} \times \frac{dh}{dt}$.	

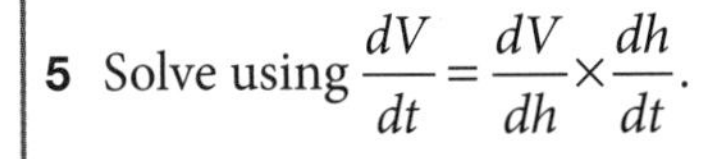

Using CAS 5: Implicit differentiation p. 241

p. 241

MATCHED EXAMPLE 18 Implicit differentiation

a Find $\frac{dy}{dx}$ for the equation $\frac{y}{x} - y = 2x$.

b Hence, find $\frac{dy}{dx}$ at $x = 2$.

Steps	Working
a **1** Differentiate both sides of the equation with respect to x, using the quotient rule for $\frac{y}{x}$.	
2 Collect $\frac{dy}{dx}$ terms and take out as a common factor.	
3 Make $\frac{dy}{dx}$ the subject.	
b **1** Find y for $x = 2$.	
2 Find $\frac{dy}{dx}$ at the *point* $(2, -8)$.	

TI-Nspire **ClassPad**

CHAPTER 6

VECTOR EQUATIONS

SB

p. 253

MATCHED EXAMPLE 1 Changing from vector to Cartesian form

Convert each vector equation to Cartesian form and describe the curve.

a $\underset{\sim}{r}(t) = (4t + 3)\underset{\sim}{i} + (2t^2 - 1)\underset{\sim}{j}$

b $\underset{\sim}{r}(t) = (3\sec(t) - 2)\underset{\sim}{i} + (4\tan(t) + 1)\underset{\sim}{j}$ for $0 \le t < 2\pi$

Steps	Working
a **1** Write in the parametric form.	
2 Write t in terms of x.	
3 Substitute in y and express in the standard form.	
4 Describe the curve.	
b **1** Write in the parametric form.	
2 Isolate sec (t) and tan (t).	
3 Use $\tan^2(\theta) + 1 = \sec^2(\theta)$.	
4 Substitute the expressions and simplify.	
5 Describe the curve.	

p. 254

MATCHED EXAMPLE 2 Describing a 3D curve from its vector equation

Describe the shape of the curve given by $\underset{\sim}{r}(t) = -t\,\underset{\sim}{i} + 2\sin(t)\,\underset{\sim}{j} + 3t\,\underset{\sim}{k}$.

Steps	Working
1 Isolate the trigonometric part.	
2 Identify the shape.	
3 Describe the whole vector.	
4 Describe the 3D shape.	

MATCHED EXAMPLE 3 Sketching the projection of a curve on the plane of two axes

SB p. 254

Sketch the projection of the curve $\underset{\sim}{r}=(2\cos(t)+1)\underset{\sim}{i}+(4\sin(t)-3)\underset{\sim}{j}+(t+1)\underset{\sim}{k}$ on the x–y plane.

Steps	Working
1 Write the projection.	
2 Write in the parametric form.	
3 Eliminate t.	
4 Describe the shape.	
5 Find the x-intercepts and y-intercepts.	
6 Sketch the projection.	

p. 256

MATCHED EXAMPLE 4 Vector equations of straight lines

Find the vector equation and parametric form of lines through

a (1, 3) and (2, 1) **b** (1, 0, 2) and (−1, −1, 2)

Steps	Working
a **1** Find the displacement vector. Write the vector equation. Write the parametric form.	
b **1** Find the displacement vector. **2** Write the vector equation. **3** Write the parametric form.	

MATCHED EXAMPLE 5	Cartesian equations of straight lines
Find the Cartesian equation of lines through **a** (2, −3) and (−1, 1) **b** (−1, 1, 2) and (2, −1, 0)	

p. 257

6

Steps	Working
a 1 Find the displacement vector.	
2 Write the Cartesian equation.	
3 For 2D cases, it is usual to write in the standard form.	
b 1 Find the displacement vector.	
2 Write the Cartesian equation.	

p. 257

MATCHED EXAMPLE 6 Vector equation of a segment

a Find the vector equation of the line segment between $P\ (2, 3, -2)$ and $Q\ (1, 2, 1)$.

b Which of the points $(3, 1, 5)$, $(2, -1, 2)$, $(1, 2.5, 3.5)$ and $(1.5, 2.5, -0.5)$ are on the line segment?

Steps	Working
a 1 Find the displacement vector.	
2 Write vector equation.	
b 1 Discard the points with a value outside the range.	
2 Find the value of t for the x-values of the remainder.	
3 Test the values of t.	
4 Write the conclusion.	

MATCHED EXAMPLE 7	Equation of a line from given information

The vectors $12\underset{\sim}{i}$, $4\underset{\sim}{j}$ and $3\underset{\sim}{k}$ are adjacent edges of a rectangular prism meeting at the point P (−2, 1, 5) that is closest to the origin. Find the equation of the diagonal through this vertex (−2, 1, 5) and the point Q furthest from it.

SB p. 258

Steps	Working
1 Sketch the prism.	
2 Find the furthest point.	
3 Find the displacement vector.	
4 Write the equation of the line *PQ*.	

p. 258

MATCHED EXAMPLE 8 Points on a plane

The plane Γ_1 contains position vectors $\underset{\sim}{a} = (1, -3, 2)$ and $\underset{\sim}{b} = (2, -1, 1)$. Determine whether the following points are on the plane.

a (4, −5, 4) **b** (10, −10, 8)

Steps	Working
a **1** Write as a combination.	
2 Simplify.	
3 Separate the components and solve.	
4 Check the values.	
5 Write the conclusion.	
b **1** Write as a combination.	
2 Simplify.	
3 Separate components and solve.	
4 Check the values.	
5 Write the conclusion.	

MATCHED EXAMPLE 9 Normal to a plane

SB p. 261

The plane Γ_1 contains position vectors $\underset{\sim}{a} = 2\underset{\sim}{i} + 5\underset{\sim}{j} - 3\underset{\sim}{k}$ and $\underset{\sim}{b} = -\underset{\sim}{i} + \underset{\sim}{j} - 2\underset{\sim}{k}$. Find a normal to Γ_1, and a normal unit vector.

Steps	Working
1 Find the vector product.	
2 Find the unit vector.	
3 Write the answer.	

p. 262

MATCHED EXAMPLE 10 Normal to a plane using three points

The points A (2, 2, 2), B (1, −2, 3) and C (1, −1, 4) are on the plane Γ_1. Find a normal to the plane.

Steps	Working
1 Find $\overrightarrow{AB}$ and $\overrightarrow{AC}$.	
2 Find the vector product.	
3 Factorise if possible.	
4 Write the answer.	

Using CAS 1: Normal to a plane with three points p. 262

MATCHED EXAMPLE 11 Equation of a plane from normal and point

p. 264

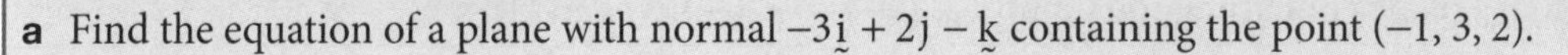

a Find the equation of a plane with normal $-3\underset{\sim}{i} + 2\underset{\sim}{j} - \underset{\sim}{k}$ containing the point (−1, 3, 2).

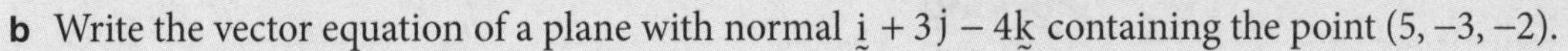

b Write the vector equation of a plane with normal $\underset{\sim}{i} + 3\underset{\sim}{j} - 4\underset{\sim}{k}$ containing the point (5, −3, −2).

Steps		Working
a	**1** Write the general formula.	
	2 Substitute values and simplify.	
	3 Write the answer.	
b	**1** Write the general formula.	
	2 Substitute values and simplify.	
	3 Write the answer.	

p. 265

MATCHED EXAMPLE 12 Equation of a plane from three points

Find the equation of a plane containing the points (2, 1, –4), (–2, –2, 3) and (1, 4, 3).

Steps	Working
1 Find two displacement vectors in the plane.	
2 Find the vector product to obtain a normal.	
3 Simplify it.	
4 Write the formula.	
5 Substitute and simplify.	
6 Write the answer.	

p. 265

6

MATCHED EXAMPLE 13 Equation of a plane from intersecting lines

Find the equation of the plane formed by the lines $\underset{\sim}{r}_1(t) = -\underset{\sim}{i} - \underset{\sim}{j} - 3\underset{\sim}{k} + t(2\underset{\sim}{i} + \underset{\sim}{j} + \underset{\sim}{k})$ and $\underset{\sim}{r}_2(s) = -3\underset{\sim}{i} + 8\underset{\sim}{j} - 2\underset{\sim}{k} + s(\underset{\sim}{i} + 3\underset{\sim}{j} + \underset{\sim}{k})$.

Steps	Working
1 State the nature of the lines.	
2 Write in the condensed form with different parameter symbols.	
3 Choose the simplest pairs of expressions and solve.	
4 Write the vectors.	
5 State the intersection.	
6 Find the vector product of the directions of the lines.	
7 Write the simplest normal.	
8 Write the formula.	
9 Substitute and simplify.	
10 Write the answer.	

p. 266

MATCHED EXAMPLE 14 Equation of a plane from parallel lines

Find the equation of a plane formed by the lines $\underset{\sim}{r}_1(t) = -\underset{\sim}{i} + \underset{\sim}{k} + t(6\underset{\sim}{i} - 2\underset{\sim}{j} + 3\underset{\sim}{k})$ and $\underset{\sim}{r}_2(s) = 4\underset{\sim}{i} - 3\underset{\sim}{j} + \underset{\sim}{k} + s(6\underset{\sim}{i} - 2\underset{\sim}{j} + 3\underset{\sim}{k})$.

Steps	Working
1 State the relationship between the lines.	
2 Choose an easy point on each line.	
3 Find the direction of the line.	
4 Find the vector product with the direction vector of the parallel lines.	
5 Write the formula.	
6 Substitute one of the points and simplify.	
7 Write the answer.	

MATCHED EXAMPLE 15	Check if a point is on a plane
Determine whether each of the following points is on the plane $3x - y - z = 15$. **a** $(3, -2, 6)$ **b** $(2, -3, -6)$ **c** $(1, 2, -4)$	
Steps	**Working**
a **1** Substitute the point in the equation and calculate the answer. **2** Write the result.	
b **1** Substitute the point in the equation and calculate the answer. **2** Write the result.	
c **1** Substitute the point in the equation and calculate the answer. **2** Write the result.	

p. 267

p. 267

MATCHED EXAMPLE 16 Intersection of planes

Find the line of intersection of the planes $3x - 2y + 4z = 2$ and $x + 2y - 2z = 4$.

Steps	Working
1 Choose the simplest variable to eliminate.	
2 Eliminate the variable.	
3 Express one variable in terms of the other.	
4 Express the first variable in terms of the same variable.	
5 Choose an expression for t in terms of the same variable to eliminate fractions.	
6 Write the other variables in terms of t.	
7 Write in vector form.	
8 Write the answer.	

INTEGRATION

p. 278

MATCHED EXAMPLE 1	Integration of polynomial function
Find the anti-derivative of $3x^5 - \frac{5}{x^5} + 2$.	
Steps	**Working**
1 Express all the parts of the polynomial in the form x^n.	
2 Anti-differentiate each term.	
3 Simplify, expressing the answer with positive indices.	

p. 279

MATCHED EXAMPLE 2	Integrating exponential functions
Find an indefinite integral of $2e^{5x}+\frac{7}{x}$.	
Steps	**Working**
Use the formulas: $\int e^{ax}\,dx=\frac{1}{a}e^{ax}+c$ and $\int \frac{1}{x}\,dx=\log_e\lvert x\rvert+c$.	
The '+ c' is not required because the question asks for 'an integral' not 'the integral'.	

p. 279

MATCHED EXAMPLE 3	Integrations that produce logarithmic functions
Find the anti-derivative of $\frac{5}{7x-3}$.	
Steps	**Working**
Use the formula: $\int \frac{1}{ax+b}\,dx=\frac{1}{a}\log_e\lvert ax+b\rvert+c$.	

MATCHED EXAMPLE 4	Integrating $(ax + b)^n$

p. 279

7

Find $\int 5(2x+1)^3\, dx$.

Steps	Working
Use the formula: $\int (ax+b)^n\, dx = \frac{(ax+b)^{n+1}}{a(n+1)} + c$.	

MATCHED EXAMPLE 5	Integrating trigonometric functions

SB

p. 279

Find $\int \sin(3x) - \cos(5x)\, dx$.

Steps	Working
Use the formulas: $\int \sin(ax)\, dx = -\frac{1}{a}\cos(ax) + c$ and $\int \cos(ax)\, dx = \frac{1}{a}\sin(ax) + c$.	

p. 280

Using CAS 1: Finding indefinite integrals p. 280

Using CAS 1: Finding definite integrals p. 280

MATCHED EXAMPLE 6 Evaluating a definite integral

Evaluate $\int_0^3 \frac{1}{2x+3}\, dx$.

Steps	Working
1 Write the anti-derivative in square brackets without the constant '*c*'. Write the limits on the right-hand bracket.	
2 Find $F(3) - F(0)$.	

MATCHED EXAMPLE 7 Integrating $\frac{1}{\sqrt{a^2 - x^2}}$

SB p. 283

Find $\int \frac{1}{\sqrt{49-x^2}}\,dx$.

Steps	Working
1 Find the value of a.	
2 Use the formula $\int \frac{1}{\sqrt{a^2-x^2}}\,dx = \sin^{-1}\left(\frac{x}{a}\right)+c$ to find the integral.	

p. 283

MATCHED EXAMPLE 8 Integrating $\frac{-1}{\sqrt{a^2 - x^2}}$

Find $\int \frac{-2}{\sqrt{4-25x^2}}\,dx$.

Steps	Working
1 Change the integrand into the form $\frac{-1}{\sqrt{a^2-x^2}}$.	
2 Find the value of a.	
3 Use the formula $\int \frac{-1}{\sqrt{a^2-x^2}}\,dx = \cos^{-1}\left(\frac{x}{a}\right)+c$ to find the anti-derivative.	

MATCHED EXAMPLE 9 Integrating $\frac{a}{a^2+x^2}$

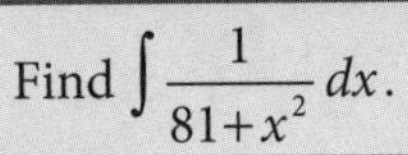

Find $\int \frac{1}{81+x^2}\,dx$.

Steps	Working
1 Find the value of a.	
2 Transform the integrand into the form $\frac{a}{a^2+x^2}$. Multiply the integral by $\frac{1}{9}\times 9$.	
3 Use the formula $\int \frac{a}{a^2+x^2}\,dx = \tan^{-1}\left(\frac{x}{a}\right)+c$ to find the anti-derivative.	

SB p. 284

p. 284

MATCHED EXAMPLE 10 Integrating $\dfrac{1}{a^2+b^2x^2}$

Find $\int \dfrac{1}{25+16x^2}\,dx$.

Steps	Working
1 Take out a factor of 16 in the denominator.	
2 Find the value of a.	
3 Transform the integrand into the form $\dfrac{a}{a^2+x^2}$. Multiply the integral by $\dfrac{1}{\frac{5}{4}}\times\dfrac{5}{4}$.	
4 Use the formula $\int \dfrac{a}{a^2+x^2}\,dx = \tan^{-1}\left(\dfrac{x}{a}\right)+c$ to find the anti-derivative.	

MATCHED EXAMPLE 11	Integration involving completing the square
Find $\int \frac{1}{x^2+6x+15}\,dx$.	
Steps	**Working**
1 Complete the square in the denominator.	
2 Find the value of a.	
3 Transform the integrand into the form $\frac{a}{a^2+x^2}$. Multiply the integral by $\frac{1}{4}\times 4$.	
4 Use the formula: $\int \frac{a}{a^2+x^2}\,dx = \tan^{-1}\left(\frac{x}{a}\right)+c$.	

SB

p. 285

7

p. 285

MATCHED EXAMPLE 12 Definite integrals that involve inverse trigonometric functions 1

Find $\int_0^3 \frac{2}{\sqrt{12-x^2}}\,dx$.

Steps	Working
1 Find the value of a.	
2 Use the formula: $\int \frac{1}{\sqrt{a^2-x^2}}\,dx = \sin^{-1}\left(\frac{x}{a}\right)$.	
3 Evaluate the definite integral.	

9780170464079

MATCHED EXAMPLE 13	Definite integrals that involve inverse trigonometric functions 2

Find $\int_{-4}^{4} \frac{1}{48+x^2} dx$.

SB

p. 286

Steps	Working
1 Find the value of a.	
2 Transform the integrand into the form $\frac{a}{a^2+x^2}$. Multiply the integral by $\frac{1}{4\sqrt{3}} \times 4\sqrt{3}$.	
3 Use the formula: $\int \frac{a}{a^2+x^2} dx = \tan^{-1}\left(\frac{x}{a}\right)$.	
4 Evaluate the definite integral. The range of arctan (x) is $\left(-\frac{\pi}{2}, \frac{\pi}{2}\right)$; therefore, $\tan^{-1}\left(\frac{-1}{\sqrt{3}}\right) = -\frac{\pi}{6}$.	

p. 288

MATCHED EXAMPLE 14 Integration by substitution involving polynomial functions 1

Find $\int (6x^2+1)\sqrt{2x^3+x}\,dx$.

Steps	Working
1 Choose a substitution for u in terms of x, which is inside the more complex function.	
2 Find $\frac{du}{dx}$.	
3 Substitute u and $\frac{du}{dx}$ into the integral.	
4 Integrate with respect to u and then substitute $u = 2x^3 + x$ to express the answer in terms of x.	

MATCHED EXAMPLE 15	Integration by substitution involving polynomial functions 2
Find $\int 9x^2(x^3+4)^4\,dx$.	
Steps	**Working**
1 Choose a substitution for u in terms of x, which is inside the more complex function.	
2 Find $\frac{du}{dx}$ and write $9x^2$ in terms of $\frac{du}{dx}$.	
3 Substitute u and $3\frac{du}{dx}$ into the integral.	
4 Integrate with respect to u and then substitute $u = x^3 + 4$ to express the answer in terms of x.	

p. 289

p. 289

MATCHED EXAMPLE 16	Integration by substitution involving sine and cosine functions
Find $\int \sin(x)\cos^5(x)\,dx$.	
Steps	**Working**
1 Choose a substitution for u.	
2 Find $\frac{du}{dx}$.	
3 Substitute u and $\frac{du}{dx}$ into the integral.	
4 Integrate with respect to u and then substitute $u = \cos(x)$ to express the answer in terms of x.	

MATCHED EXAMPLE 17 Integration by linear substitution

Find $\int x\sqrt{3x-5}\,dx$.

Steps	Working
1 Choose a substitution for u in terms of x, which is inside the more complex function.	
2 Transpose to express x in terms of u.	
3 Find $\frac{du}{dx}$ and write an equation in the form $k\frac{du}{dx}=1$.	
4 Substitute for u, x and 1 in the integral.	
5 Integrate with respect to u and then substitute $u = 3x - 5$ to express the answer in terms of x.	

SB

p. 290

7

p. 290

MATCHED EXAMPLE 18 Substitution with definite integrals

Find the value of $\int_0^{\frac{\pi}{2}} \sin(x)\cos^3(x)\,dx$.

Steps	Working
1 Choose a substitution for u.	
2 Find $\frac{du}{dx}$.	
3 Convert the limits of integration from x to u by substituting into $u = \cos(x)$.	**Lower limit** **Upper limit**
4 Substitute u, $\frac{du}{dx}$ and the limits into the integral. Every part of the integral MUST be in terms of the substituted variable u, including limits.	
5 Integrate with respect to u and then substitute the limits of integration into the anti-derivative.	

MATCHED EXAMPLE 19	**Linear substitution with definite integrals**
Find the value of $\int_6^9 \frac{x-3}{\sqrt{x-5}}\,dx$.	

SB

p. 291

7

Steps	**Working**
1 Choose a substitution for u.	
2 Transpose to express x in terms of u.	
3 Find $\frac{du}{dx}$.	
4 Convert the limits from x values to u values.	
5 Substitute u, $\frac{du}{dx}$ and the limits into the integral.	
6 Integrate and evaluate.	

p. 294

MATCHED EXAMPLE 20 Integrating an odd power of a sine function

Find $\int \sin^5(x)\,dx$.

Steps	Working
1 Write the integrand as $\sin(x) \times$ an even power of $\sin(x)$.	
2 Substitute $\sin^2(x) = 1 - \cos^2(x)$.	
3 Choose a substitution for u.	
4 Find $\frac{du}{dx}$.	
5 Substitute for u and $-\frac{du}{dx}$ in the integral.	
6 Integrate with respect to u and then substitute $u = \cos(x)$ to express the answer in terms of x.	

MATCHED EXAMPLE 21	Integrating the square of a sine function
Find $\int \sin^2(5x)\,dx$.	
Steps	**Working**
1 Substitute $\sin^2(5x) = \frac{1}{2}[1 - \cos(10x)]$.	
2 Integrate.	

p. 295

7

p. 295

MATCHED EXAMPLE 22	Integrating the square of a cosine function
Find $\int 3\cos^2(x)\,dx$.	
Steps	**Working**
1 Substitute $\cos^2(x)=\frac{1}{2}[1+\cos(2x)]$.	
2 Integrate.	

p. 296

MATCHED EXAMPLE 23	Integration using trigonometric identities 1
Find $\int \tan^2(3x)+2\,dx$.	
Steps	**Working**
Substitute $\sec^2(x)=1+\tan^2(x)$.	

MATCHED EXAMPLE 24 Integration using trigonometric identities 2

SB p. 296

Find $\int 2\sin^2(x)\sin(2x)\,dx$.

Steps	Working
1 Use the double-angle rule for cos (2x). $\cos(2x) = 1 - 2\sin^2(x)$ $1 - \cos(2x) = 2\sin^2(x)$	
2 Use the double-angle rule for sin (2x). $\sin(2x) = 2\sin(x)\cos(x)$ $\frac{1}{2}\sin(4x) = \sin(2x)\cos(2x)$	

p. 296

MATCHED EXAMPLE 25 Integration by substitution

Find $\int_0^{\frac{\pi}{6}} \sin^5(x)\cos(x)\,dx$.

Steps	Working
1 Write $\sin^5(x)$ as $\sin(x) \times$ an even power of $\sin(x)$.	
2 Substitute $\sin^2(x) = 1 - \cos^2(x)$.	
3 Choose a substitution for u.	
4 Find $\frac{du}{dx}$.	
5 Convert the limits of integration from x values to u values.	
6 Substitute u, $-\frac{du}{dx}$ and the limits into the integral. Note that the new limits are not in ascending order like the original limits are.	
7 Integrate and evaluate.	

MATCHED EXAMPLE 26 Integration by partial fractions with two linear factors

Find $\int \frac{x+6}{(x+2)(x-2)}\,dx$.

Steps	Working
1 Write $\frac{x+6}{(x+2)(x-2)}$ as the sum of two partial fractions.	
2 Write with a single denominator and equate the numerators.	
3 Solve for A and B by equating coefficients.	
4 Substitute the values of A and B into the partial fractions	
5 Use the partial fractions to find the integral.	

p. 299

7

Using CAS 3: Partial fractions p. 300

p. 300

MATCHED EXAMPLE 27 Definite integration by partial fractions

Find the exact value of $\int_7^8 \frac{2x-14}{x^2-8x+15}\,dx$.

Steps	Working
1 Factorise the denominator.	
2 Write as partial fractions.	
3 Write with a single denominator and equate the numerators.	
4 An alternative to equating coefficients is to substitute values of x that eliminate A or B.	
5 Rewrite the integral into partial fractions.	
6 Evaluate the definite integral and simplify using logarithm laws.	

MATCHED EXAMPLE 28 Integrating partial fractions with repeated linear factors

p. 301

Find $\int \frac{x-5}{(x+1)^2}\,dx$.

Steps	Working
1 Write $\frac{x-5}{(x+1)^2}$ as the sum of two partial fractions.	
2 Write with a single denominator and equate the numerators.	
3 Solve for A and B by equating coefficients.	
4 Integrate the partial fractions.	

p. 303

MATCHED EXAMPLE 29 Integration by recognition

Find the derivative of $\sin^2 x$, and hence find $\int \sin x \cos x \, dx$.

Steps	Working
1 Find the derivative.	
2 Write the corresponding integral.	
3 Find the given integral.	

MATCHED EXAMPLE 30	Integration by splitting the numerator
Find $\int \frac{x-3}{\sqrt{4-x^2}}\,dx$.	
Steps	**Working**
1 Split the numerator to create two integrals.	
2 The second integral produces $\sin^{-1}\left(\frac{x}{2}\right)$. Choose a suitable substitution for u in the first integral.	
3 Find the integrals.	

SB

p. 304

7

p. 306

MATCHED EXAMPLE 31 Integration by parts

Find $\int xe^{2x}\,dx$.

Steps	Working
1 Identify $f(x)$ and $g'(x)$, where $f'(x)$ is simpler than $f(x)$. Enter these functions in the grid.	
2 Find $f'(x)$ and $g(x)$, and enter them in the grid.	
3 Substitute in the formula $\int f(x)g'(x)\,dx$ $= f(x)g(x) - \int f'(x)g(x)\,dx$	

MATCHED EXAMPLE 32 Two-step integration by parts

Find $\int x^2 \sin x \, dx$.

Steps	Working
1 Identify $f(x)$ and $g'(x)$, where $f(x)$ is the function with the simpler derivative. Enter these functions in the grid.	
2 Find $f'(x)$ and $g(x)$, and enter them in the grid.	
3 Substitute in the formula: $\int f(x)g'(x)\,dx = f(x)g(x) - \int f'(x)g(x)\,dx$	
4 We need to integrate by parts again to find $\int x \cos x \, dx$. Repeat the process.	
5 Substitute this result for the integral in step 3.	

SB

p. 306

7

p. 307

MATCHED EXAMPLE 33 Integration by parts involving inverse trigonometric functions

Find $\int \arcsin(x)\, dx$.

Steps	Working
1 Let $g'(x) = 1$. Enter $f(x)$ and $g'(x)$ in the functions grid.	
2 Find $f'(x)$ and $g(x)$, and enter them in the grid.	
3 Substitute into the integration by parts formula.	
4 Find $\int \frac{x}{1-x^2}\, dx$ by substitution.	
5 Substitute the integral found into the integral equation in step 3.	

MATCHED EXAMPLE 34 Integration by parts involving exponential functions

p. 307

7

Find $\int e^x \sin(x)\,dx$.

Steps	Working
1 Let $f(x) = e^x$ and $g'(x) = \sin(x)$. Enter $f(x)$ and $g'(x)$ in the functions grid.	
2 Find $f'(x)$ and $g(x)$, and enter them in the grid.	
3 Substitute into the integration by parts formula.	
4 We need to integrate by parts again to find $\int e^x \cos(x)\,dx$. Repeat the process.	
5 Substitute into the integral equation in step 3 and solve.	

CHAPTER 8 AREAS AND VOLUMES OF INTEGRATION

p. 318

MATCHED EXAMPLE 1 Finding the area between a curve and the x-axis

Find the area between the curve $f(x)=\frac{36}{9+x^2}$ and the x-axis from $x=-3$ to $x=3$.

Steps	Working
1 Graph the function and identify the area required.	
2 Write a definite integral that represents the area and simplify the integral by symmetry.	
3 Find the anti-derivative and evaluate the definite integral.	

MATCHED EXAMPLE 2 Finding the area bounded by a curve and the x-axis

p. 318

Find the area bounded by the function $f(x) = \cos(\pi x)$ and the x-axis between $x = -\frac{5}{2}$ and $x = -\frac{1}{2}$.

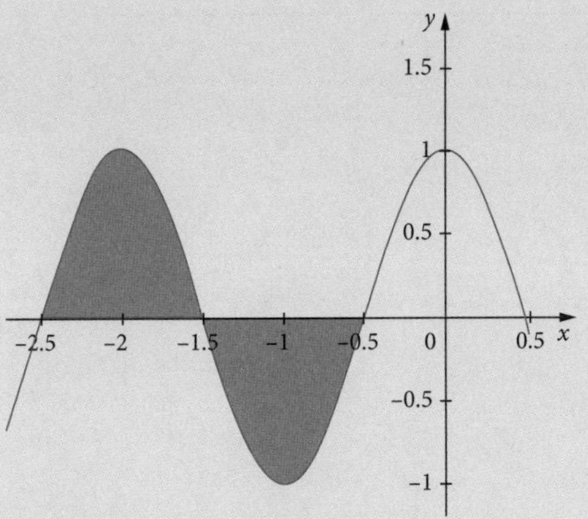

Steps	Working
1 Write a definite integral that represents the area.	
2 Find the anti-derivative and evaluate the definite integral.	

Using CAS 1: Finding the area bounded by a curve and the x-axis p. 319

SB

p. 321

MATCHED EXAMPLE 3 Finding the area between two curves

Find the area bounded by the functions $f(x) = -x^2 + 4$ and $g(x) = (x-2)^2$.

Steps	Working
1 Find the x-coordinates of the intersection points between the two functions.	
2 Sketch the graphs of f and g on the same axes.	
3 Area between two curves $= \int (\text{top curve} - \text{bottom curve})\, dx$	

SB

Using CAS 2: Finding the area between two curves p.321

p. 323

8

MATCHED EXAMPLE 4 Applying the area between a curve and the y-axis

Find the area bound by the function $f(x) = \cos^{-1}(x)$ and the lines $x = -1$ and $y = \frac{\pi}{2}$.

Steps	Working
1 Graph the function and identify the area required.	
2 Write a definite integral for the required area.	
3 We must find the area between $f(x)$ and the y-axis and subtract it from the area of a rectangle. Transpose to make x the subject and convert the upper and lower limits from x values to y values.	
4 Find the area between f and the y-axis.	
5 Subtract this area from the area of the rectangle.	

p. 328

MATCHED EXAMPLE 5 Sketching a possible anti-derivative function

Sketch the graph of a possible anti-derivative for $y = f(x)$.

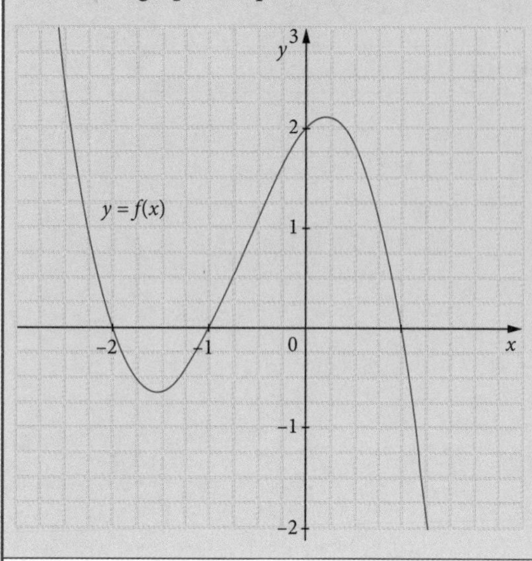

Steps	Working
1 $y = f(x)$ has three x-intercepts at $x = -1$, $x = 1$ and $x = -2$. They represent the three stationary points with a gradient of zero. Draw horizontal lines below each x-intercept, for the slope graph.	
2 Draw an increasing line in the section of the slope graph where $f(x)$ is above the x-axis and a decreasing line in the section where $f(x)$ is below the x-axis.	

3 Move the horizontal lines on the slope graph to make the slope graph continuous.

4 Draw a smooth curve from the slope graph and translate the graph upwards to any position on the axes above.

The value of the constant is not known, so any vertical translation can be used.

8

p. 329

MATCHED EXAMPLE 6 Sketching the anti-derivative function

Sketch the graph of the anti-derivative for $y = f(x)$, where $\int f(x)dx = F(x)$, given that $F(0) = 1$.

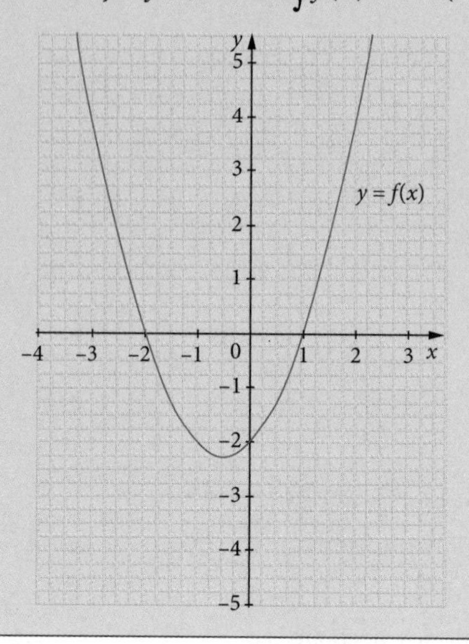

Steps	Working
1 $y = f(x)$ has two x-intercepts at 1 and −2. They represent the two stationary points with a gradient of zero. There are two stationary points with a gradient of zero at $x = 1$ and $x = -2$. Draw a horizontal line below each x-intercept. Add the slope lines below the graph obtained from the gradient function $f(x)$.	
2 As $F(0) = 1$, the graph of the anti-derivative $F(x)$ passes through the point (0, 1). Draw a smooth curve from the slope graph and translate the graph upwards so that it passes through the point (0, 1).	

MATCHED EXAMPLE 7 Volumes of solids of revolution

p. 332

8

Find the volume generated when the area between the function $y = x + 1$ and the x-axis from $x = 2$ to $x = 6$ is rotated about the x-axis.

Steps	Working
1 Graph the function and identify the area required.	
2 Rotate the curve in a circular motion about the x-axis to create a solid of revolution.	
3 Write the formula for the volume of a solid of revolution about the x-axis. Substitute $y = x + 1$ using $x = 2$ and $x = 6$ as the lower and upper limits of integration.	
4 Evaluate the definite integral.	

p. 334

MATCHED EXAMPLE 8 Finding the volume of a solid of revolution about the *y*-axis

Find the volume generated when the area between the function $y = \sin^{-1}\left(\frac{x}{2}\right)$, $x \in [0, 2]$ and the y-axis is rotated about the y-axis.

Steps	Working
1 Graph the function and identify the area required.	
2 Rotate the curve in a circular motion about the y-axis to create a solid of revolution.	
3 Write the formula for the volume of a solid of revolution about the y-axis.	
4 Transpose the equation to make x the subject.	
5 Substitute into the volume formula using $y = 0$ and $y = \frac{\pi}{2}$ as the lower and upper limits of integration.	
6 Use the double-angle rule $\sin^2(x) = \frac{1}{2}[1 - \cos(2x)]$ to find the anti-derivative.	

MATCHED EXAMPLE 9 Revolving regions bounded by two curves

p. 335

8

The area enclosed by $y = x^2$, $y = 2x$ and the line $x = 2$ is rotated about the x-axis to form a solid of revolution. Find the volume of the solid.

Steps	Working
1 Graph the function and identify the area required.	
2 The function, when rotated about the x-axis, will form a hollow solid. Write a definite integral to represent this volume.	
3 Integrate and evaluate.	

p. 342

MATCHED EXAMPLE 10 The arc length of a curve

Find the arc length of the curve $y = \frac{2}{3}x^{\frac{3}{2}}$ between $x = 0$ and $x = 2$.

Steps	**Working**
1 Find $\frac{dy}{dx}$ and $\left(\frac{dy}{dx}\right)^2$.	
2 Use the arc length formula.	
4 Find the anti-derivative using the formula $\int (ax+b)^n \, dx = \frac{(ax+b)^{n+1}}{a(n+1)} + c$.	
5 Evaluate the definite integral.	

Using CAS 3: Calculating the length of a curve p. 343

MATCHED EXAMPLE 11 Arc length in parametric form

SB p. 344

Find the arc length for the curve defined by the parametric function
$x = 1 + 2\cos(t), y = 2 - 2\sin(t), t \in [0, 3]$.

Steps	Working
1 Find $\frac{dx}{dt}$ and $\frac{dy}{dt}$.	
2 Substitute into the formula $l = \int_{t_1}^{t_2} \sqrt{\left(\frac{dx}{dt}\right)^2 + \left(\frac{dy}{dt}\right)^2}\, dt$.	
3 Use the identity $\sin^2(x) + \cos^2(x) = 1$ to find the anti-derivative.	
4 Evaluate the definite integral.	

p. 348

MATCHED EXAMPLE 12 Surface area of a solid of revolution

Find the surface area generated by rotating the function $f:[-1,1]\rightarrow R, f(x)=\sqrt{4-x^2}$ about the x-axis.

Steps	Working
1 Find $f'(x)$.	
2 Substitute into the formula: $S=\int_a^b 2\pi f(x)\sqrt{1+\left(f'(x)\right)^2}\,dx$	
3 Integrate	

MATCHED EXAMPLE 13 Surface area of a solid of revolution about the y-axis

p. 349

8

Find the surface area generated when the function $y = \sqrt[3]{x}$ on the interval $[0, 8]$ is rotated about the y-axis.

Steps	Working
1 Find x as a function of y, $\frac{dy}{dx}$ and the new limits of integration.	
2 Substitute into the formula: $S = \int_c^d 2\pi x\sqrt{1+\left(\frac{dx}{dy}\right)^2}\, dy$.	
3 Integrate by substitution.	

p. 350

MATCHED EXAMPLE 14 Surface area of a solid of revolution in parametric form

Find the surface area generated by rotating the function with parametric equations

$x(t) = \sin^2(t)$, $y(t) = \cos^2(t)$, $t \in \left[0, \frac{\pi}{2}\right]$ about the x-axis.

Steps	Working
1 Find $x'(t)$ and $y'(t)$.	
2 Substitute into the formula: $S = \int_{t_1}^{t_2} 2\pi y(t)\sqrt{(x'(t))^2 + (y'(t))^2}\, dt$.	
3 Integrate by substitution.	

9780170464079

CHAPTER

DIFFERENTIAL EQUATIONS

9

MATCHED EXAMPLE 1 Finding the differential equation

p. 362

Express the information as a differential equation using the variables given.

a The rate of change of the surface area, A cm^2, of a cube with respect to its length, L cm, is directly proportional to its length.

b The rate of change of the distance, x m, travelled by a particle with respect to time, t hours, is inversely proportional to the square of the distance.

c The rate of increase in the population P with respect to time, t minutes, of a certain organism in an ecosystem with an initial population of 1000 is proportional to the product of its current population P and the decrease in its population.

Steps	Working
a **1** Express the rate of change in terms of the given variables.	
2 'Proportional' means the rate of change is the product of a constant with some function of the independent variable.	
b **1** Express the rate of change in terms of the given variables.	
2 'Inversely proportional' means the rate of change is the ratio of a constant and some function of the independent variable.	
c **1** Express the rate of change in terms of the given variables.	
2 Express the rate of change in terms of the given variables.	

p. 363

MATCHED EXAMPLE 2 Verifying a solution to a differential equation

Verify that $y = \tan(x)$ is a solution to the second-order differential equation $\frac{d^2y}{dx^2} - 2y = 2y^3$.

Steps	Working
1 Use CAS to find an expression for $\frac{d^2y}{dx^2}$ in terms of y.	
2 Substitute for $\frac{d^2y}{dx^2}$ in the left-hand side and simplify.	
3 State the conclusion.	

MATCHED EXAMPLE 3 Verifying a solution to find the constant

Determine the values of the constant m in $y = e^{mx}$ that satisfy $\frac{d^2y}{dx^2} - 10\frac{dy}{dx} + 16y = 0$.

Steps	Working
1 Use $y = e^{mx}$ to find $\frac{dy}{dx}$ and $\frac{d^2y}{dx^2}$.	
2 Substitute the information found in the differential equation.	
3 Take out a common factor and simplify.	
4 Solve the quadratic equation by factorising.	

SB
p. 363

Using CAS 1: Verifying solutions to differential equations
p. 364

p. 368

MATCHED EXAMPLE 4 General solution to a first-order, first-degree differential equation

Find the general solution to each differential equation.

a $\frac{dy}{dx} = \frac{2}{4+x^2}$ **b** $\frac{dy}{dt} = 2t^3 - 4t + 3$ **c** $\frac{dp}{dy} = 2\cos(y) + 3$

Steps	**Working**
a Use $\int \frac{1}{a^2+x^2}\,dx = \frac{1}{a}\tan^{-1}\left(\frac{x}{a}\right) + c$.	
b Find the anti-derivative and simplify.	
c Find the anti-derivative and simplify. Write in the usual order.	

MATCHED EXAMPLE 5	**Particular solution to a first-order, first-degree differential equation**

Find the solution to each differential equation.

a $\frac{dy}{dx} = e^{2x} - 4x$ if $y(0) = 2$ **b** $\frac{dy}{dx} = (3x-5)^{\frac{3}{2}}$ if $y(2) = \frac{1}{5}$

c $\frac{dy}{dx} = \sin^2(x)$ if $y\left(\frac{\pi}{2}\right) = \frac{\pi}{2}$

SB p. 369

Steps	Working
a Find the anti-derivative and simplify.	
Substitute $y(0) = 2$ to find c.	
Write the answer.	
b Find the anti-derivative using the formula $\int (ax+b)^n \, dx = \frac{(ax+b)^{n+1}}{a(n+1)} + c$	
Substitute $y(2) = \frac{1}{5}$ to find c.	
Write the answer.	
c Use the formula $\cos(2x) = 1 - 2\sin^2(x)$ to rewrite the integral.	
State the anti-derivative.	
Substitute $y\left(\frac{\pi}{2}\right) = \frac{\pi}{2}$ to find c.	
Write the answer.	

9

Using CAS 2: Solving $\frac{dy}{dx} = f(x)$ p. 370

p. 374

MATCHED EXAMPLE 6 Anti-differentiating twice

Determine the general solution to y for $\frac{d^2y}{dx^2} = \cos(2x) - \sin(x)$.

Steps	Working
1 Anti-differentiate to obtain $\frac{dy}{dx}$, and include the first constant of integration.	
2 Anti-differentiate $\frac{dy}{dx}$ to obtain the equation for y. Include the second constant of integration.	

MATCHED EXAMPLE 7 Finding the velocity from acceleration

The acceleration of a moving particle is given by $a = 2t + 6\ \text{ms}^{-2}$, where t is in seconds. If the particle travels a distance of 4 m at $t = 6$ s, find the velocity and the distance travelled by the particle at 10 s.

SB

p. 375

Steps	Working
1 The anti-derivative of velocity gives the displacement.	
Write the differential equation and integrate it.	
Simplify.	
Use $v(0) = 0$ to find c.	
2 Write the equation and substitute $t = 10$ to find the velocity at 10 s.	
3 Find the distance function by integrating the velocity.	
Use $x(6) = 4$ to find k.	
Write the equation.	
4 Substitute $t = 10$ to find the distance at 10 s.	
Write the answer.	

9

Using CAS 3: Solving $\frac{d^2y}{dx^2} = f(x)$ p. 376

p. 380

MATCHED EXAMPLE 8 Solving $\frac{dy}{dx} = f(y)$

Solve each differential equation.

a $\frac{dy}{dx} = \frac{1}{y^2}$ **b** $\frac{dx}{dt} = \frac{x+1}{t-2}$ **c** $\frac{dQ}{dt} = t^2,\ Q(0) = 0$

Steps	Working
a **1** Transpose so that the same variables are together and show as an integral.	
2 Anti-differentiate both sides.	
3 Solve for y and add a constant of integration.	
b **1** Transpose so that the same variables are together and show as an integral.	
2 Anti-differentiate both sides.	
3 Solve for x and add a constant of integration.	
c **1** Transpose so that the same variables are together and show as an integral.	
2 Anti-differentiate both sides.	
3 Calculate the value of the constant.	
4 Substitute the constant into the general solution and solve for the dependent variable.	

MATCHED EXAMPLE 9 Newton's Law of Cooling

p. 381

A cup of oil is heated to 84°C and then begins to cool according to Newton's law of cooling. The surrounding temperature is 24°C, and the oil takes 8 minutes to drop to 54°C.

Find, to the nearest minute, how long it will take for the temperature of the oil to drop to 44°C.

Steps	Working
1 Write the required differential equation, including the known information.	
2 Write the integral and anti-differentiate.	
3 Calculate the value of the constant using the initial temperature.	
4 Calculate the value of the remaining constant using the temperature at 8 minutes. Do not round your value of k so that you can reuse it without losing accuracy. Store it on your calculator or write it down with many decimal places.	
5 Find the time required for the oil's temperature to drop to the given value.	
6 State the answer.	

p. 382

MATCHED EXAMPLE 10	Find the age of a substance using the rate of decay differential equation

The rate of decay of radioactive material depends on the amount of material present at the time. The half-life is the time taken for half of a given amount of material to decay. The half-life of uranium-238 is 4468 million years. A rock sample contains 25 g of uranium-238, and it is estimated that 100 g of uranium-238 was present at the time the rock was formed. Use a differential equation to find the age when the rock was formed. Give your answer to the nearest million year.

Steps	Working
1 Choose the variables and write the equation. Decreasing gives a negative rate of change. Place M on the left.	
2 Integrate both sides with respect to time and make M the subject. M_0 is the initial value of M. To find k, use the half-life.	
3 Write the equation for M.	
4 Substitute $M = 25$ to find t, the time when the rock was formed.	
5 Write the answer.	

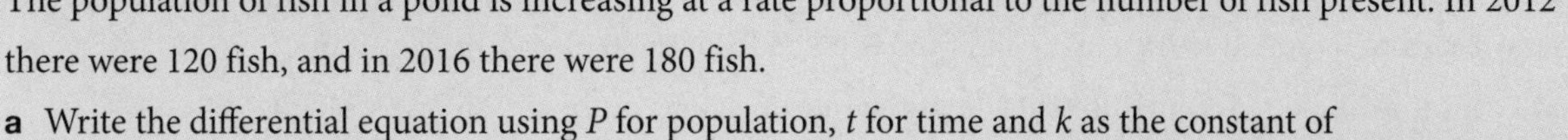

MATCHED EXAMPLE 11 Apply the growth model to predict the population

SB p. 383

The population of fish in a pond is increasing at a rate proportional to the number of fish present. In 2012 there were 120 fish, and in 2016 there were 180 fish.

a Write the differential equation using P for population, t for time and k as the constant of proportionality.

b Solve the differential equation.

c Calculate the predicted population for 2023.

Steps	Working
a Use the general form $\frac{dy}{dx} \propto f(y)$ with the variables given.	
b **1** Anti-differentiate.	
2 Find the value of the constant of integration. Take 2012 as the start ($t = 0$).	
3 Determine the value of the constant of proportionality.	
4 Transpose to make P the subject.	
c **1** Substitute the value for t in the equation.	
2 State the answer.	

p. 384

MATCHED EXAMPLE 12 Solving the inflow/outflow differential equation

An irrigation tank contains 100 litres of water, in which 5 kg of fertiliser is dissolved. More fertiliser of ratio 1 kg/litre is added to the tank, and liquid is taken from the tank at 2 litres/minute.

a Form the differential equation that shows the rate of change of the fertiliser dissolving.

b Solve the differential equation.

c Correct to one decimal place, find the amount of fertiliser in the tank after 20 minutes.

d Sketch the graph of the amount of fertiliser, M against t, and describe what happens to the amount of fertiliser in the tank in the limiting case.

Steps	Working
a 1 Determine the inflow rate and the outflow rate.	
2 Use rate = inflow rate – outflow rate.	
b 1 Separate the variables and anti-differentiate.	
2 Evaluate the constant.	
3 Write the equation with M as the subject.	
c Use the equation to find the amount of fertiliser for the given time.	
d 1 Use CAS to draw the graph for positive values of t.	
2 Describe the behaviour of the graph for large values of t.	

MATCHED EXAMPLE 13 Apply the logistic model to a population of rabbits

p. 385

An island is currently home to 250 rabbits, and it is estimated that it can sustain a population of 1500. After 3 years, it is estimated that the number of rabbits will be 550.

a State the logistic equation for the number of rabbits, P, at time t years.

b Solve the equation for P.

c Use the logistic equation to find, correct to two decimal places, how long it will take for the rabbit population to reach 800.

d Sketch the graph of P against t, and explain the significance of the value of the carrying capacity.

Steps	Working
a Substitute the known information into $\frac{dP}{dt} = kP\left(1 - \frac{P}{K}\right)$.	
b **1** Group the variables P and t, use partial fractions and write in integral form.	
2 Anti-differentiate.	
3 Evaluate the constant of integration.	
4 Evaluate the constant of proportionality.	
5 Write the equation with P as the subject.	

9

c Substitute the given value of P, and solve for t algebraically or use SOLVE on CAS.

TI-Nspire

ClassPad

d 1 Use CAS to sketch the graph.

2 Describe the graph in terms of intercepts and asymptotes.

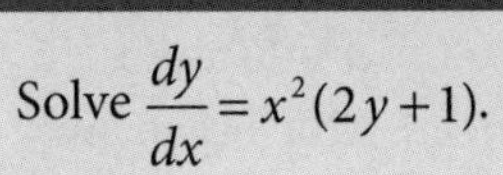

MATCHED EXAMPLE 14 Separation of variables

SB p. 394

9

Solve $\frac{dy}{dx} = x^2(2y+1)$.

Steps	Working
1 Separate the variables.	
2 Integrate both sides.	
It is easiest if the coefficient of y is 1.	
Simplify.	
Make y the subject.	

p. 395

MATCHED EXAMPLE 15 Separation of variables problem

The rate at which microbial organisms grow in a petri dish of water is proportional to the product of their number and the time they were cultured in the lab. Four hours after the culture, there are 1200 organisms; after another 4 hours, there are 2400 organisms. How long will it take for the number to increase to 3000?

Steps	Working
1 Choose variables and write the rate equation for n.	
2 Separate variables and integrate both sides.	
Make n the subject.	
3 Rename constants to simplify.	
4 Use $n(4) = 1200$ and $n(8) = 2400$ to find k.	
5 Divide the second equation by the first.	
6 Write the equation for n.	
Use $n(4) = 1200$ to find A.	
7 Substitute $n = 3000$ to find t.	
8 Write the answer.	

 9780170464079

MATCHED EXAMPLE 16 Sketching a slope field

a Sketch the slope field for $\frac{dy}{dx} = \frac{x}{4}$ for $-3 \le y \le 3$.

b Describe what type of function y must be.

p. 399

Steps	Working
a 1 Complete a table of $\frac{dy}{dx}$ values.	
2 Draw the grid, and decide if the slopes will be the same vertically or horizontally.	
3 Sketch the slope lines.	
b 1 Connect the pattern of arrows to show the general shape of the anti-derivative function.	
2 Describe the type of curve.	

Using CAS 4: Slope fields p. 400

p. 409

MATCHED EXAMPLE 17 Euler's method

Use a step size of $h = 0.2$ with the initial condition (0, 2) to find the first three iterations for the solution to $\frac{dy}{dx} = x^2$.

Steps	Working
1 Use the initial condition to find the coordinates for the first iteration by substituting into Euler's formula.	
2 Repeat the process to work out the remaining iterations.	

MATCHED EXAMPLE 18 Comparing Euler's method to the exact answer

a Use calculus to solve $\frac{dy}{dx} = x^3$ given that when $x = 0$, $y = 1$.

b Use a step size of $h = 0.4$ with the initial condition (0, 1) to obtain a series of approximations in the interval $0 \le x \le 2$ to solve $\frac{dy}{dx} = x^3$. State answers correct to three decimal places.

c Find the percentage error, to the nearest whole value, between the last approximation and the exact value.

p. 409

Steps	Working
a Anti-differentiate the function, and find the constant of integration using the initial condition.	
b **1** Write down the known information, and determine the number of iterations.	
2 Write the general form for the solution.	
3 Calculate each iteration up to the required number.	
c **1** Find the exact value	
2 Calculate the percentage error.	

Using CAS 5: Euler's method p. 409

CHAPTER 10 KINEMATICS

SB

p. 426

MATCHED EXAMPLE 1 Average speed and velocity

The graph shows the position x cm of a particle at time t seconds.

a Illustrate the motion of the particle as a position–time line in the interval [0, 4].

b Calculate the average speed in the interval [0, 4].

c Determine the average velocity in the interval [0, 4].

Give answers to two decimal places.

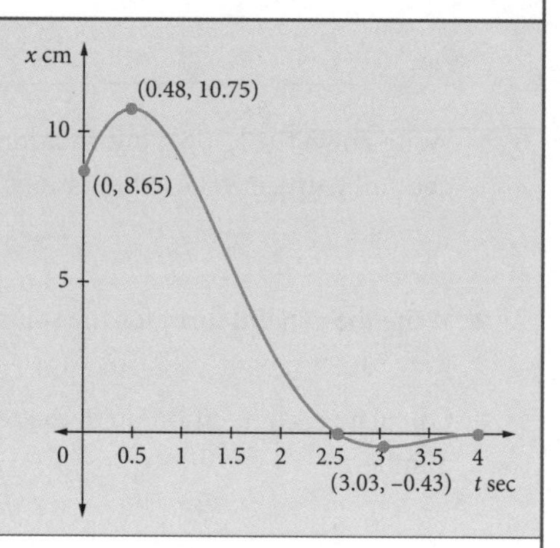

Steps	Working
a 1 Determine the initial displacement.	
2 Describe the motion for each part of the graph.	
3 Draw the position–time line.	
b average speed $= \dfrac{\text{total distance}}{\text{total time}}$	

c average velocity = $\frac{\text{displacement}}{\text{time taken}}$

10

p. 427

MATCHED EXAMPLE 2 Average speed, velocity and direction

An aircraft flies in the direction 045°T for 2 hours 50 minutes at a constant speed of 360 km/h and then flies due east for a further 4 hours at a constant speed of 500 km/h.

a Find the average speed of the plane.

b Calculate the magnitude of the average velocity.

c In which direction is the average velocity vector?

Give answers to two decimal places.

Steps	Working
a 1 Show the information as a vector diagram.	
2 Calculate the average speed.	
b 1 Find the displacement using the cosine rule.	
2 Calculate the magnitude of the average velocity.	
c 1 Use the sine rule to calculate angle $\theta°$ and use it to find the direction of the velocity vector.	
2 State the direction.	

9780170464079

MATCHED EXAMPLE 3 Finding velocity and acceleration from displacement

p. 429

10

In relation to a fixed point, the position x cm of a particle moving horizontally at time t seconds is described by the function $x = \frac{2}{3}t^3 - 7t^2 + 20t$.

a Determine when the velocity of the particle is zero and its position at that time.

b Find at what time the acceleration is zero and the particle's velocity at this time.

c Calculate the distance that the particle will travel during the first 5 seconds.

Steps	Working
a 1 Differentiate the equation to obtain the velocity function.	
2 Let the equation for velocity be zero and solve for t.	
3 Determine the position using these t values.	
b 1 Differentiate the velocity equation to obtain the acceleration function.	
2 Use the equation and the given value for acceleration to find the value for t.	
3 Find the particle's velocity by substituting the value for t.	
c Use the graph of the position function to find the distance in each section. Turning points occur when the velocity is zero.	

p. 430

MATCHED EXAMPLE 4 Velocity and position from acceleration

The acceleration a m/s^2 of a particle at time t seconds is given by the function $a = 12t - 4$.

a Obtain the equation for the velocity v m/s of the particle given that $v = 40$ when $t = 3$.

b Find the equation for the position x m of the particle given that $x = 5$ when $t = 2$.

c Find the exact value of the acceleration when the particle's velocity is 0.

Steps	Working
a Anti-differentiate the equation for acceleration with respect to t to obtain the velocity function. Then find the value of the constant of integration.	
b Anti-differentiate the velocity equation with respect to t to obtain the position function and find the value of the constant of integration.	
c **1** Find the value of t when the velocity is zero. **2** Substitute $t = 1$ into the equation for acceleration as time cannot be negative.	

9780170464079

SB
p. 431

10

MATCHED EXAMPLE 5 Position from acceleration

The acceleration a cm s^{-2} of an object moving in a straight line with positive velocity v cm/s when x cm from the origin is given by $a = 3(x - 1)^3$. Find the position, x cm, of the object at time t seconds given that $v = \sqrt{\frac{3}{2}}$ when $x = 0$, and when $t = 0$, $x = 0$.

Steps	Working
1 Decide which form of acceleration can be used.	
2 Anti-differentiate with respect to x and find the value of the constant of integration.	
3 Write v as $\frac{dx}{dt}$.	
4 Anti-differentiate and find the value of the constant of integration.	
5 Solve for x.	

SB

p. 431

MATCHED EXAMPLE 6 Velocity as a function of position

The velocity, v km/h, of a particle x km from a fixed point is $v = e^{-kx}$, where k is a constant.

a Determine the value of k if the acceleration is 5 km h^{-2} when the velocity is 1 km/h.

b Obtain an expression for the position of the particle at any time given that its initial position is zero.

Steps	Working
a **1** Use the equation for velocity to find an expression for acceleration.	
2 Evaluate k using the given information.	
b **1** Anti-differentiate velocity to obtain displacement.	
2 Find the constant.	
3 Write displacement as a function of time.	

SB

Using CAS 1: Displacement, velocity and acceleration p. 432

MATCHED EXAMPLE 7	Velocity–time graph 1
A stone 15 m above the ground begins to fall. It hits the ground after 3 s and rebounds with one-third of its impact speed. Taking the gravitational acceleration as 10 m/s^2, show the information as a velocity–time graph for $0 \le t \le 3.5$.	

SB p. 439

Steps	Working
1 Take 'down' to be positive velocity. Determine the velocity on impact.	
2 Work out the time taken to reach maximum height after the rebound. Note that the velocity is now negative.	
3 Show this information on the graph.	

p. 440

MATCHED EXAMPLE 8 Velocity–time graph 2

A toy car starting from rest uniformly accelerates and takes 3 seconds to reach a velocity of 7 cm/s. It takes another 3 seconds to change its velocity from 7 cm/s to −2 cm/s, which it maintains for a further 5 seconds. The toy car takes another 3 seconds to uniformly change its velocity from −2 cm/s to 7 cm/s.

a Show the velocity–time graph that describes the motion of the toy car.

b Find the total distance travelled.

c Determine the toy car's displacement.

Steps	Working
a 1 Determine the shape of the velocity–time graph for each section.	
2 Sketch the velocity–time graph.	
b 1 Calculate the area between the graph and the horizontal axis.	
2 State the distance travelled by the toy car.	
c Calculate the signed area between the graph and the horizontal axis.	

MATCHED EXAMPLE 9 Velocity–time graph of two moving objects

p. 441

An electric bike passes Amy's house at a speed of 36 km/h, which is maintained. Amy immediately leaves in a van from the house in pursuit and uniformly accelerates to reach a speed of 54 km/h in 12 seconds, which is maintained until she catches up to the bike adjacent to a church.

a Show the information as a velocity–time graph.

b Determine how long the electric bike and van take to reach the church.

c How far is the church from Amy's house?

Steps	Working
a **1** Convert units from km/h to m/s.	
2 Determine the shape of the graph for the electric bike.	
3 Determine the shape of the graph for the van.	
4 Sketch the velocity–time graph.	
b **1** Calculate the area under the graph for the electric bike and van.	
2 Find the value of t by equating and solving the two expressions for area.	
c **1** Use the value of t found to calculate the distance required.	

10

p. 442

MATCHED EXAMPLE 10 Terminal velocity of a falling object

A particle initially at rest begins to fall vertically with acceleration $a = 3g - 2kv$, where v m s^{-1} is its velocity, $g = 9.8$ is a constant and k is another constant.

a Express the velocity, v, as a function of time, t.

b Find the value of the constant, k, correct to two decimal places given that $v(15) = 50$.

c State the **terminal velocity**, giving your answer correct to the nearest integer.

d How far (correct to one decimal place) has the object fallen in the time interval [3, 8]?

Steps	Working
a **1** Write the acceleration in the form $\frac{dv}{dt}$.	
2 Use separation of variables and integrate, including the constant of integration.	
3 Determine the value of the constant of integration.	
4 Write the velocity in terms of time.	
b Find the other constant from the information given.	
c Terminal velocity is the value v approaches as t approaches infinity.	

d **1** Distance fallen is the integral of the velocity function in the given interval.

2 State the answer to the required accuracy.

10

p. 443

MATCHED EXAMPLE 11 Velocity–time graph of van and jeep

A van is parked 30 m from a street corner at the start of a long straight road. The van starts moving away from it just as a jeep enters the straight at the corner. The speed of the jeep is given by $v = 9 + 2e^{0.25t}$ m/s and the speed of the van is $v = 3t$ m/s.

a Sketch a velocity–time graph for $t = 0$ to 20 s.

b Use your CAS calculator to find when and where the jeep passes the van, correct to the nearest metre.

Steps	Working
a **1** State nature of van v–t graph.	
2 State nature of jeep graph.	
3 Sketch the graphs.	
b **1** Find the distance of the van from the corner.	
2 Find the distance of the jeep from the corner.	
3 Use CAS to solve the two equations simultaneously: van distance = jeep distance	

TI-Nspire

ClassPad

4 Write the time.

5 Find the distance.

6 Write the answer.

p. 451

MATCHED EXAMPLE 12 Constant retardation

A vehicle reduces its speed from 55 m/s to 40 m/s in 15 s. Assuming constant retardation, find how much distance the vehicle covers when it comes to a complete stop.

Steps	Working
1 List the known values.	
2 Decide the formula to use first.	
3 Find the distance for the vehicle's speed to go from 55 m/s to 40 m/s.	
4 Find the distance for the vehicle's speed to go from 40 m/s to 0 m/s.	
5 State the total distance covered.	

MATCHED EXAMPLE 13 Motion of two objects moving simultaneously

p. 451

In a kart race, a Drakan Spyder passes the starting point with speed 360 km/h, which it maintains for the remainder of the race. Five seconds later, an Ariel Atom 4 starts from rest and uniformly accelerates over a distance of 800 m to reach a speed of 450 km/h, which it then maintains for the remainder of the race.

a Determine when the Ariel Atom 4 catches up with the Drakan Spyder.

b The Ariel Atom 4 finishes the race after 70 seconds and then uniformly decelerates and stops in 20 seconds. To the nearest metre, what is its total distance from start to when it comes to a stop?

Steps	Working
a 1 Convert km/h to m/s.	
2 Work out the time taken for the Ariel Atom 4 to reach its maximum speed.	
3 Set up the equation for the distance travelled by each kart.	
4 The distance is the same when both karts meet.	
b 1 Find the length of the racetrack.	
2 Calculate the distance that the Atom Ariel 4 travels from its maximum speed to when it stops.	
3 State the total distance that the Atom Ariel 4 travels.	

p. 452

MATCHED EXAMPLE 14	Upward motion under gravity

A pendulum rises vertically with an initial velocity of 20 m/s. Show that by taking $g = 10$ m/s^2, the time it takes for the pendulum to reach half its maximum height is $2-\sqrt{2}$ seconds.

Steps	Working
1 Find the time taken to reach the maximum height. Take upward motion to be positive.	
2 Calculate the maximum height reached by the pendulum.	
3 Determine the time taken to reach half the maximum height.	

MATCHED EXAMPLE 15 Two bodies moving in opposite direction

Ben is paragliding, and the paraglider wings are rising at 5 m/s. When he is 600 m above ground, one shoe comes off from his foot. Neglecting air resistance and using $g = 9.8$ m/s^2, find how long it will take for the shoe to hit the ground. Give your answer correct to two decimal places.

SB

p. 453

Steps	Working
1 The shoe will continue to rise for a time. Find the time taken to reach its maximum height.	
2 Find the total time for the shoe to return to its initial height.	
3 Calculate the time taken to reach the ground from its initial height. The acceleration is negative because the motion is now opposite to the original direction of motion.	
4 Find the total time taken.	

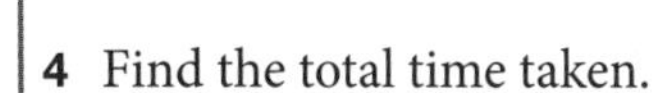

10

SB

Using CAS 2: Displacement, velocity and gravitational acceleration p. 454

p. 454

MATCHED EXAMPLE 16 Downward motion and differential equations

From a hot-air balloon, a ball is thrown vertically downward with initial speed 1 m/s. As it falls, it experiences retardation that is proportional to its speed, v m/s.

Take the gravitational acceleration as $g = 9.8$ m/s^2.

a State the differential equation involving v and time t that describes the situation.

b Solve the equation in part **a** to express v in terms of t, given that the velocity of the ball after 10 s is 64 m/s.

c What velocity does the ball approach as the value of t becomes large? Give the answer to the nearest integer.

d Determine how far, to the nearest integer, the ball has dropped after 10 seconds.

Steps	Working
a Choose the form of acceleration that has the required variables. The resultant acceleration consists of the positive gravitational acceleration and the negative retardation.	
b **1** Write the integrands in the right form and anti-differentiate. **2** Find the constant of integration, substitute into the equation and make v the subject of the equation. **3** Determine the value of k using CAS.	

c Examine the value of v as t becomes large (approaches infinity).

d Distance travelled is the integral of the velocity equation in the given time interval.

p. 458

MATCHED EXAMPLE 17 Uniform increase in acceleration

A van initially travelling at 15 m/s uniformly increases its acceleration from 5 m s^{-2} to 12 m s^{-2} in 25 seconds. Find the van's speed and distance travelled after 25 seconds.

Steps	Working
1 Uniform increase in acceleration is a linear function of the form $\frac{dv}{dt} = mt + c$. Find the gradient m and intercept c.	
2 Anti-differentiate to find the velocity function.	
3 Find the velocity at 25 seconds.	
4 Anti-differentiate to find the displacement function.	
5 Find the displacement by finding the area under the curve.	

MATCHED EXAMPLE 18 Constant and uniform acceleration increase

p. 459

A body with initial speed 3 cm/s is moving with acceleration 4 cm s^{-2}, which it maintains for 5 seconds. It then takes 12 seconds to uniformly increase its acceleration to 12 cm s^{-2}. The body's acceleration is then uniformly reduced to zero during a further 8 seconds.

a Sketch the acceleration–time graph.

b Sketch the velocity–time graph.

c Calculate the total distance the body has travelled.

Steps	Working
a 1 Describe the acceleration in each required section.	
2 Show the information as an acceleration–time graph.	

b **1** Obtain expressions by anti-differentiation that describe the velocity in each required section.

2 Show the information as a velocity–time graph.

c To find the total distance, use calculus or CAS to find the area under each section of the velocity–time graph.

SB

Using CAS 3: Find the time taken for a given distance p. 460

VECTOR CALCULUS

MATCHED EXAMPLE 1	Position vectors
a Find the position of a particle with the position vector function $\underset{\sim}{r}(t)=(t^2+1)\underset{\sim}{i}+2t\underset{\sim}{j}$ at $t=0$, $t=1$ and $t=-1$. **b** Find the distance of the particle from the origin at $t=2$.	
Steps	**Working**
a Substitute $t=0$, $t=1$ and $t=-1$.	
b Use distance $=\lvert\underset{\sim}{r}(t)\rvert=\sqrt{(x(t))^2+(y(t))^2}$.	

p. 476

p. 477

MATCHED EXAMPLE 2 Cartesian equations

a Find the Cartesian equation of the vector function $\underset{\sim}{r}(t) = (5t)\,\underset{\sim}{i} - (2t^2)\,\underset{\sim}{j}$ for $t \geq 0$.

b Give the domain and range of the Cartesian equation and sketch its graph.

Steps	Working
a 1 Use the vector function to write x and y as functions of t.	
2 Eliminate t.	
b 1 Consider $x = 5t$ and $y = -2t^2$ for $t \geq 0$.	
2 Cartesian equation: $y = -\frac{2x^2}{25}$	

MATCHED EXAMPLE 3 Cartesian graphs

Sketch the Cartesian graph of the vector function $\underset{\sim}{r}(t) = \frac{1}{2}\sin(t)\,\underset{\sim}{i} + \frac{1}{2}\cos(t)\underset{\sim}{j},\ t \geq 0$.

Steps	Working
1 Write two equations for x and y, and rearrange them into a possible trigonometric identity.	
2 Use the trigonometric identity $\sin^2(t) + \cos^2(t) = 1$.	
3 Check the domain and range for $t \geq 0$.	
4 Sketch the graph of $x^2 + y^2 = \frac{1}{4}$, a circle with the centre at the origin and radius $\frac{1}{2}$.	

SB p. 478

p. 481

MATCHED EXAMPLE 4 Velocity and acceleration

a Find the velocity and acceleration of an object with the position vector $\underset{\sim}{r}(t) = t^2\underset{\sim}{i} + \cos(t)\underset{\sim}{j} + e^{2t}\underset{\sim}{k}$.

b Hence find the velocity and acceleration at $t = \frac{\pi}{2}$.

Steps	Working
a **1** Find $\dot{\underset{\sim}{r}}(t)$. **2** Find $\ddot{\underset{\sim}{r}}(t)$.	
b Substitute $t = \frac{\pi}{2}$ into $\dot{\underset{\sim}{r}}(t)$ and $\ddot{\underset{\sim}{r}}(t)$.	

MATCHED EXAMPLE 5	Speed
a Find the speed function from the velocity vector $\dot{\underset{\sim}{r}}(t) = 3t\underset{\sim}{i} + \sin(t)\underset{\sim}{j} + 2e^t\underset{\sim}{k}$. **b** Hence, find the speed at $t = 0$.	

p. 482

Steps	Working
a Find the magnitude of $\dot{\underset{\sim}{r}}(t)$.	
b Substitute $t = 0$ into $\lvert\dot{\underset{\sim}{r}}(t)\rvert$.	

11

p. 482

MATCHED EXAMPLE 6	Velocity, acceleration, displacement
The acceleration in m/s^2 of a particle is $\ddot{\underset{\sim}{r}}(t) = 2t\underset{\sim}{i} + \sin(t)\underset{\sim}{j} + \cos(2t)\underset{\sim}{k}$. Find the velocity and displacement of the particle, given that initially the particle has a velocity of $2\underset{\sim}{j}$ m/s and a position of $3\underset{\sim}{k}$ metres.	
Steps	**Working**
1 Integrate to find $\dot{\underset{\sim}{r}}(t)$.	
2 Substitute the initial velocity to find $\underset{\sim}{c}$. At $t = 0$, $\dot{\underset{\sim}{r}}(0) = 2\underset{\sim}{j}$.	
3 Find the expression for $\dot{\underset{\sim}{r}}(t)$	
4 Integrate to find $\underset{\sim}{r}(t)$.	
5 Substitute the initial displacement to find $\underset{\sim}{d}$. At $t = 0$, $\underset{\sim}{r}(0) = 3\underset{\sim}{k}$.	
6 Find the expression for $\underset{\sim}{r}(t)$.	

MATCHED EXAMPLE 7	Vectors in motion 1

SB p. 486

11

The motion of two particles is given by the vector functions $\underset{\sim}{r}_1(t)=(2t)\underset{\sim}{i}+(t^2)\underset{\sim}{j}$ and $\underset{\sim}{r}_2(t)=(4t-6)\underset{\sim}{i}+(2t+3)\underset{\sim}{j}$, where $t \geq 0$.

a Find the point at which the particles collide.

b Find the point(s) at which their paths cross.

Steps	**Working**
a **1** Equate the $\underset{\sim}{i}$ and $\underset{\sim}{j}$ coefficients, and solve for t.	
2 Substitute the same value of t into the vector functions.	
3 Conclude whether the particles collide or cross.	
b **1** Write the two vectors using a different variable, s, for time in $\underset{\sim}{r}_2$.	
2 Equate the $\underset{\sim}{i}$ and $\underset{\sim}{j}$ coefficients, and solve simultaneously for t and s.	
TI-Nspire	ClassPad
3 Substitute the values into the vector functions and write the answer.	
4 Conclude whether the particles collide or cross.	

p. 487

MATCHED EXAMPLE 8 Vectors in motion 2

The position vector of a particle at time t is given by $\underset{\sim}{r}(t) = (5\sin(3t))\underset{\sim}{i} + (5\cos(3t))\underset{\sim}{j}$, where $t \geq 0$.

a Find the Cartesian equation of the path of the particle.

b Sketch the graph of the path.

Steps	Working
a **1** Write parametric equations for x and y	
2 Identify the trigonometric identity to be used.	
3 Find the Cartesian equation.	
4 Identify where $t = 0$ and the direction of the path.	
b **1** Identify where the Cartesian graph exists.	

9780170464079

2 Sketch the Cartesian graph $x^2 + y^2 = 25$.

The particle starts at (0, 5) and travels around the circle in a clockwise direction.

11

p. 488

MATCHED EXAMPLE 9 Vectors in motion 3

The position, $\underset{\sim}{r}(t)$, of a projectile at time t seconds is given by $\underset{\sim}{r}(t)=(20t)\underset{\sim}{i}+(56t-4.9t^2)\underset{\sim}{j}$, where $t \geq 0$. The object is initially on the ground, and the distance is in metres. Find

a the time taken to return to the ground.

b the maximum height reached.

c the initial speed of the object.

Give all answers correct to 2 decimal places.

Steps	**Working**
a The projectile is at ground level when the vertical component of its position equals zero.	
b The maximum height reached is when the vertical component of velocity equals zero. Differentiate the vertical component of displacement and solve for zero	
c **1** Find the velocity vector of the particle at $t = 0$. **2** Find the speed at $t = 0$.	

RANDOM VARIABLES AND HYPOTHESIS TESTING

MATCHED EXAMPLE 1 Finding the mean of $aX + b$ for a discrete probability distribution

SB p. 502

The probability distribution of a discrete random variable X is shown.

x	0	1	2	3
Pr $(X = x)$	0.2	0.3	0.3	0.2

a Find the mean of X.

b Find $E(2X + 4)$.

Steps	Working
a Use the formula $E(X) = \sum x \cdot p(x)$. Rewrite the table above using columns. Add a column with the heading '$x \times p(x)$', and calculate the product of the x values and their probabilities. The total of the $x \times p(x)$ column is the expected value of X or $E(X)$.	
b Find $E(2X + 4)$ by substituting into the formula $E(aX + b) = a\,E(X) + b$ with $a = 2$ and $b = 4$.	

p. 503

MATCHED EXAMPLE 2 Finding the mean of $aX + b$ for a continuous probability distribution

The probability density function for a continuous random variable X is given by

$$f(x) = \begin{cases} \frac{3}{4}(2x - x^2), & 0 \le x \le 2 \\ 0, & \text{elsewhere} \end{cases}$$

Find

a $E(X)$

b $E(3X + 1)$

Steps	Working
a **1** Use the formula $\mu = E(X) = \int_{-\infty}^{\infty} x f(x) dx$ and simplify.	
2 Evaluate the integral.	
b Substitute $a = 3$ and $b = 1$ into the formula $E(aX + b) = a\,E(X) + b$.	

SB

p. 504

12

MATCHED EXAMPLE 3 Finding the variance of aX + b for a continuous probability density function

The probability density function for a continuous random variable X is given by

$$f(x)=\begin{cases}\dfrac{3}{4}(2x-x^2), & 0\le x\le 2\\ 0, & \text{elsewhere}\end{cases}$$

Find the variance and hence find $\mathrm{Var}(3X+1)$.

Steps	Working
1 Write the formula for the variance and calculate $\mathrm{E}(X)$ and $\mathrm{E}(X^2)$. The mean μ or $\mathrm{E}(X)$ was calculated in Matched example 2.	
2 $\mathrm{E}(X^2)=\int_{-\infty}^{\infty} x^2\, f(x)\,dx$	
3 Substitute in the variance formula from Step 1.	
4 Substitute into the formula $\mathrm{Var}(aX+b)=a^2\,\mathrm{Var}(X)$.	

p. 505

MATCHED EXAMPLE 4 Finding the mean and variance of the sum of two variables

At Dream Oven bakery, the time taken to make a tiered wedding cake, X hours, is a continuous random variable with a mean time of 8 hours and a variance of 5 hours, and the time taken to assemble and frost the cake, Y hours, is a continuous random variable with a mean time of 3 hours and a variance of 1 hour. Find the mean and variance of the time taken to finish making the cake.

Steps	Working
1 The total time taken to finish the cake is the sum of the time to make the cake and the time to assemble and frost the cake.	
2 Substitute $a = 1$ and $b = 1$ into the formula $\text{E}(aX + bY) = a\,\text{E}(X) + b\text{E}(Y)$ to find the mean time.	
3 Substitute $a = 1$ and $b = 1$ into the formula $\text{Var}(aX + bY) = a^2\,\text{Var}(X) + b^2\,\text{Var}(Y)$ to find the variance.	

MATCHED EXAMPLE 5	Finding E($aX + bY$) and Var($aX + bY$)
Two independent random variables X and Y have means of 50 and 30 and variances of 9 and 6 respectively. If $Z = 6X + 3Y$, find the mean and variance of Z.	
Steps	**Working**
1 To find the mean of Z, use the formula $\mathrm{E}(aX + bY) = a\,\mathrm{E}(X) + b\,\mathrm{E}(Y)$	
2 To find the variance of Z, use the formula $\mathrm{Var}(aX + bY) = a^2\,\mathrm{Var}(X) + b^2\,\mathrm{Var}(Y)$	

p. 505

12

p. 506

MATCHED EXAMPLE 6	Finding the mean and variance of the sum of n variables of type X and m variables of type Y

Almonds used in a homemade chocolate shop have a mean mass of 500 g with a standard deviation of 150 g, and walnuts used in the same chocolate shop have a mean mass of 350 g with a standard deviation of 95 g.

The masses of the almonds are independent of the masses of the walnuts.

Find the mean mass and standard deviation, in grams, of a set of six of these almonds and four of these walnuts.

Steps	Working
a **1** Find the mean using the formula $E(x_1 + x_2 + \ldots + x_n + y_1 + y_2 \ldots + y_m)$ $= n\,E(X) + mE(Y)$, where $n = 6$ and $m = 4$	
2 Find the variance using the formula $\text{Var}(x_1 + x_2 + x_3 \ldots + x_n + y_1 + y_2 + y_3 \ldots + y_m)$ $= n\,\text{Var}(X) + m\text{Var}(Y)$, where $n = 6$ and $m = 4$	
3 Find the standard deviation.	

MATCHED EXAMPLE 7 The sum of two normally distributed random variables

The weights of carrots harvested in a farm are normally distributed with a mean of 68 kg and a standard deviation of 4 kg, and the weights of potatoes harvested from the same farm are normally distributed with a mean of 64 kg and a standard deviation of 3 kg.

If the weights are independent, find

a The mean and the standard deviation of the combined weights of a carrot and a potato.

b A carrot and a potato are randomly selected from the harvest. Find the probability, correct to three decimal places, that the carrot weighs more than the potato.

p. 509

Steps	Working
a **1** Let X = weight of a carrot and Y = weight of a potato.	
2 Use the formula $E(aX + bY) = a\mu_X + b\mu_Y$ to find the mean of the combined weights.	
3 Use the formula $Var(aX + bY) = a^2(\sigma_X)^2 + b^2(\sigma_Y)^2$ to find the variance of the combined weights.	
b **1** Write a probability equation.	
2 Find the mean and variance of $X - Y$ by using the formulas for mean and variance mentioned in part **a**.	
3 Find the standard deviation.	
4 Calculate the normal probability using CAS.	

Using CAS 1: Finding normal probabilities p. 510

Using CAS 2: Finding the inverse probability for a normal distribution p. 511

p. 511

MATCHED EXAMPLE 8 The sum of *n* items selected from a normal distribution

The volume of ice-cream dispensed by a machine into cups varies normally with a mean of 125 mL and a standard deviation of 3 mL. The ice-cream is sold in boxes of 12 cups. Let the normal random variable X represent the total volume of ice-cream (mL) in a pack of 12 cups.

a Find the mean and standard deviation of X.

b Find $\Pr(X < 1505)$, correct to three decimal places.

c Twelve boxes of ice-cream are rejected if their total volume is less than x mL. Find the value of x correct to the nearest mL if 2% of packs are rejected.

Steps	**Working**
a 1 Find the mean using the formula $E(x_1 + x_2 + x_3 \ldots + x_n)$ $= n\,E(X)$, where $n = 12$.	
2 Find the variance and standard deviation using the formulas $\text{Var}(x_1 + x_2 + x_3 \ldots + x_n) = n\,\text{Var}(X)$ where $n = 12$ and $\text{sd}(X) = \sqrt{\text{Var}(X)}$.	
b Calculate the normal probability using CAS.	
c Write the probability equation and use the inverse cumulative normal on CAS to solve.	
TI-Nspire	**ClassPad**

Using CAS 3: The distribution of the sample means by simulation p. 515

p. 517

12

MATCHED EXAMPLE 9 **Finding probabilities from a simulated sample**

A random sample is taken from a population that is normally distributed with a mean of 75 and a standard deviation of 7. Through simulation, 100 of these random samples is generated and the results are shown in the dotplot.

Find the probability that a sample contains a mean greater than 77.

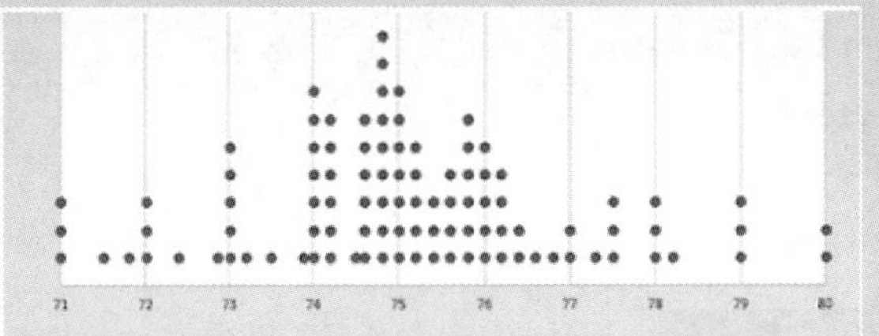

Steps	**Working**
Count the number of sample means greater than 77. The probability is a fraction of the total number of samples.	

p. 517

MATCHED EXAMPLE 10 Finding the mean of the sample means using simulation

The amount of time a person spends on their phone, in a day, is normally distributed with a mean of 360 minutes and a standard deviation of 20 minutes.

a Simulate 50 samples of size 70, and calculate the sample means for each sample. Display the results as a dotplot or histogram.

b Find the average of the sample means for the samples of size 70.

Steps	Working
a Generate the means of the samples by CAS using the randNorm command.	
TI-Nspire	ClassPad
b Calculate one-variable statistics for the data in column A. Answers may vary slightly.	

MATCHED EXAMPLE 11 The mean and standard deviation of sample means for different sample sizes

p. 518

Random samples of size n are taken from a population that is normally distributed with a mean of 120 and a standard deviation of 8. Complete 100 simulations for sample sizes of 20 and 100, and determine the mean and standard deviation of the sample means.

Steps	Working
1 Generate the means of the samples by CAS using the randNorm command. Use $n = 20$ and $n = 100$ in the formulas.	
TI-Nspire	**ClassPad**
2 Answers may vary slightly.	

12

p. 519

MATCHED EXAMPLE 12	The standard error of the sample mean
A random sample of 500 customers was taken at a mall. The time spent shopping at the mall is normally distributed with a mean time of 120 min and a standard deviation of 10 min. Find the standard error of the sample mean.	
Steps	**Working**
Use the formula $SE(\bar{x}) = \frac{\sigma}{\sqrt{n}}$	

MATCHED EXAMPLE 13 Finding a confidence interval for a normally distributed variable

SB p. 522

The time, in minutes, that readers spend at the library is found to be a normally distributed random variable with a mean of 45 minutes and a standard deviation of 8 minutes.

Find a 98% confidence interval, correct to one decimal place, for this variable.

Steps	Working
1 Determine the percentage of values in each tail.	
2 Use the inverse cumulative normal distribution to determine the quantiles of the confidence interval.	
3 Use CAS to determine the quantiles of the confidence interval.	
TI-Nspire	ClassPad

p. 524

MATCHED EXAMPLE 14 Finding the probability for the sampling distribution of the sample means.

The weight of newborn babies in a hospital is normally distributed with a mean of 3500 g and a standard deviation of 600 g. What is the probability that the mean weight of a sample of 36 babies will be less than 3800 g (correct to four decimal places)?

Steps	Working
1 Estimate the standard deviation of the sampling distribution.	
2 Since the population is normal, the sampling distribution of the sample means is also normal.	
3 Find the probability using CAS.	
TI-Nspire	**ClassPad**

MATCHED EXAMPLE 15 Finding a confidence interval for the population mean

SB

p. 524

A random sample of size 40 is taken from a population that is normally distributed with a standard deviation of 8. If the mean of the sample is 85, find a 95% confidence interval for the population mean (answer correct to 2 decimal places).

Steps	Working
1 Estimate the standard deviation of the sampling distribution.	
2 Since the population is normal, the sampling distribution of the sample means is also normal.	
3 Determine the percentage of values in each tail.	
4 Use the inverse cumulative normal distribution to determine the quantiles of the confidence interval.	
5 Use CAS to determine the quantiles of the confidence interval. Enter the sample mean as the mean of the sampling distribution.	
TI-Nspire	**ClassPad**

12

SB

Using CAS 4: The confidence interval for the population mean p. 525

p. 526

MATCHED EXAMPLE 16 Confidence interval formula

At a cocoa plantation, the weight of a random sample of 50 cocoa pods is recorded. The sample has a mean of 450 g and a standard deviation of 1 g. Find a 95% confidence interval for the mean mass of cocoa pods at the plantation (correct to three decimal places).

Steps	Working
1 Estimate the standard error for the sampling distribution using the sample standard deviation, s, as an estimate of the population standard deviation σ.	
2 Since $n \geq 30$, you can use the normal distribution.	
3 Determine the z-value used for a 95% confidence interval.	
4 Substitute into the formula for the confidence interval.	

MATCHED EXAMPLE 17	Finding the sample size n, for a given margin of error
A population has a standard deviation of 12. Determine the sample size, n, needed so that the 95% confidence level for the population mean is within 6 of the sample mean.	
Steps	**Working**
Use the margin of error formula to calculate n.	

p. 526

p. 527

MATCHED EXAMPLE 18 Finding the sample mean and size, for a given confidence interval

The standard deviation of the weights of croissants in a bakery is 1 g. From the results of a sample of n croissants, a 95% confidence interval for the mean weight of croissants was calculated to be (44.72, 45.28). Calculate the mean score and the size of this random sample.

Steps	Working
1 Calculate the sample mean by finding the median of the confidence interval.	
2 Calculate the margin of error and substitute into the formula $M = z\frac{\sigma}{\sqrt{n}}$ using $z = 1.96$ and $\sigma = 1$	
3 Solve the equation for n using CAS.	

SB

p. 532

12

MATCHED EXAMPLE 19 The null and alternative hypothesis 1

A soda company claimed that a can of soda has a mean of 100 g of sugar and a standard deviation of 10 g. An article published in a journal stated that an average can of sugar-sweetened soda consists of more sugar than what the company claims. Write the null and alternative hypotheses.

Steps	Working
1 The null hypothesis is a statement about the population that is assumed to be true.	
2 The alternative hypothesis is a statement about the population mean that contradicts the null hypothesis. The article in the journal stating that an average can of sugar-sweetened soda consists of more sugar than what the company claims indicates that the direction of the alternative hypothesis must be $\mu >$ the population mean.	

p. 532

MATCHED EXAMPLE 20 The null and alternative hypothesis 2

A bakery makes cupcakes with an average mass of 80 g. A chef who weighed 50 cupcakes believes the mean weight to be different. Write the null and alternative hypotheses.

Steps	Working
1 The null hypothesis is a statement about the population mean that is assumed to be true.	
2 The alternative hypothesis is a statement about the population mean that contradicts the null hypothesis. As the chef disputes the claim, the alternative hypothesis is $\mu \neq$ the population mean.	

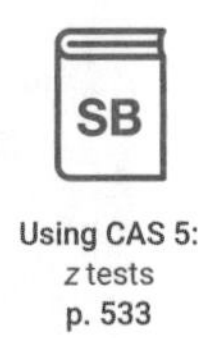

Using CAS 5: *z* tests p. 533

MATCHED EXAMPLE 21 One-tailed test

SB p. 534

A company that sells peanut butter claims that a jar of opened peanut butter has a shelf life with a mean of 85 days and a standard deviation of 5 days. A bakery that uses the peanut butter believes the shelf life to be less and tests a random sample of 60 jars and finds the mean of the sample to be 83 days. Test the hypothesis that the shelf life of peanut butter is less than what the company claims. Test at the 5% significance level.

Steps	Working
1 Write the null hypothesis.	
2 Write the alternative hypothesis.	
3 Find the p-value. Assume that H_0 is true. Find the probability of obtaining the sample mean or a value more extreme in the direction of the alternative hypothesis. As H_1 is $\mu < 85$ and the observed mean in the sample is 83, the observed value or a more extreme value is $\bar{x} < 83$.	
4 Complete a one-sample z test. The mean used is the population mean. The z test can be completed using CAS.	
TI-Nspire	**ClassPad**
Write the p-value.	

5 Make a decision about the null hypothesis for the 5% level of significance.

Reject if $p < 0.05$.

Fail to reject if $p > 0.05$.

6 Make a conclusion about the claim.

MATCHED EXAMPLE 22 Two-tailed test

A company sells oil and sets its machines up to dispense the same amount of oil in each bottle. The mean and standard deviation of the volume of oil in each bottle are 550 ml and 9 ml, respectively. A consumer group claims the volume of oil in each bottle is not 550 ml. The group then tests a sample of 50 bottles and finds the average volume to be 546 ml. Test the consumer group's claim at the 1% significance level.

p. 536

12

Steps	Working
1 Write the null hypothesis.	
2 Write the alternative hypothesis. The alternative hypothesis is that the volume of oil in a bottle is not 550 ml.	
3 Assume that H_0 is true. Find the probability of obtaining the sample mean or a value more extreme in the direction of the alternative hypothesis. As H_1 is $\mu \neq 550$, a two-tailed test is required. We calculate the probability of the observed value $\bar{x} = 546$, or a more extreme value is $(\bar{x} < 546)$.	
4 Complete a one-sample z test. The mean used is the population mean. The p-value is found using CAS. Input the data using CAS.	
TI-Nspire	ClassPad
5 Make a decision about the null hypothesis. When p-value < 0.01, the observed difference is highly significant	
6 Make a conclusion about the claim. Since the null hypothesis H_0 is rejected and the alternative hypothesis is H_1: $\mu \neq 550$, we need to decide if $\mu > 550$ or $\mu < 550$. The sample mean $\bar{x} = 546$ is less than 550; therefore, we choose $\mu < 550$.	

Using CAS 6: Finding the critical z-value p. 537

p. 538

MATCHED EXAMPLE 23 Testing a null hypothesis using a critical z-value

A company sells breakfast cereal in boxes with an advertised mean mass of 350 g. Their competitors claim that on average the boxes contain less cereal than advertised. They take a random sample of 25 containers and find the mean mass to be 347 g. The mass of cereal in all boxes is normally distributed with a standard deviation of 15 g.

a State the appropriate null and alternative hypotheses for the volume.

b The p-value for this test is given by the expression $\Pr(Z < z)$, where Z has the standard normal distribution. Find the value of z, and hence determine whether the null hypothesis should be rejected at the 0.05 level of significance.

Steps	Working
a **1** Write the null hypothesis.	
2 Write the alternative hypothesis.	
b **1** Find the value of the test statistic z. Use $z = \dfrac{\bar{x} - \mu}{\sigma_{\bar{X}}}$, where $\sigma_{\bar{X}} = \dfrac{\sigma}{\sqrt{n}}$.	
2 Compare the test statistic with the critical z-value for the significance level of 0.05.	
3 Make a decision about the null hypothesis.	

MATCHED EXAMPLE 24 Finding critical values of the sample mean

p. 538

A normally distributed population has a mean of 85 and a standard deviation of 7.

Consider the hypotheses H_0: $\mu = 85$ and H_1: $\mu \neq 85$.

a Find the critical z-values at the 0.02 significance level, correct to three decimal places.

b A random sample of size 16 is taken from this population. The random variable $\bar{X}$ represents the sampling distribution of the sample means.

Find the critical values of $\bar{X}$ at which the null hypothesis would be rejected.

Steps	**Working**
a 1 Draw the standard normal distribution, and label the critical z-values and the associated probabilities. A two-tailed test is required for $\mu \neq 85$.	
2 Use an inverse cumulative normal distribution on CAS to find z. In the two-tailed test, we reject the null hypothesis $\mu = 85$ if the test statistic is in the rejection region shaded in the standard normal distribution.	
TI-Nspire	**ClassPad**
b 1 Find the value of $\sigma_{\bar{X}}$	
2 $\bar{X}$ is the sampling distribution of the sample means of size 16 with a mean of 85 and a standard error of 1.75. The value of c corresponds to the critical z-value of 2.326. Substitute into $z = \frac{\bar{x} - \mu}{\sigma_{\bar{X}}}$. The value of d corresponds to the critical z-value of −2.326. Substitute into $z = \frac{\bar{x} - \mu}{\sigma_{\bar{X}}}$. The critical values of the sample mean can also be found using CAS by finding the 96% confidence interval with mean = 85, standard deviation = 7, sample size = 16 and confidence level = 0.96.	

p. 546

MATCHED EXAMPLE 25 Type I and II errors

The battery life of a drone is 15 minutes. The company that manufactures the drone claims that the new version of the drone has increased battery life.

a Write the null and alternative hypotheses.

b Describe a type I error and the impact in this situation.

c Describe a type II error and the impact in this situation.

Steps	Working
a Write the null and alternative hypotheses.	
b A type I error occurs when we reject the null hypothesis when it is true.	
c A type II error occurs when we fail to reject the null hypothesis when it is false.	

Answers

Worked solutions available on Nelson MindTap.

CHAPTER 1

MATCHED EXAMPLE 1

$\overrightarrow{CD} = -\underset{\sim}{a} + \underset{\sim}{b}$

MATCHED EXAMPLE 2

$\hat{\underset{\sim}{b}} = \frac{1}{\sqrt{6}}\left(2\underset{\sim}{i} - \underset{\sim}{j} - \underset{\sim}{k}\right)$

MATCHED EXAMPLE 3

a $m = \frac{1}{2}$

b $\underset{\sim}{a}$, $\underset{\sim}{c}$ and $\underset{\sim}{d}$ are **not linearly dependent.**

MATCHED EXAMPLE 4

a $\overrightarrow{PQ} = -2\underset{\sim}{i} + 3\underset{\sim}{j} + \underset{\sim}{k}$

b $\overrightarrow{RS} = 5\underset{\sim}{i} + \underset{\sim}{k}$

c $\overrightarrow{OA} = \frac{5}{\sqrt{2}}\underset{\sim}{i} + \frac{5}{\sqrt{2}}\underset{\sim}{j}$

MATCHED EXAMPLE 5

$\gamma = 74.50°$

MATCHED EXAMPLE 6

a -2

b $x = 6$

MATCHED EXAMPLE 7

$\theta = 108.91°$

MATCHED EXAMPLE 8

$\frac{13}{6}\underset{\sim}{i} + \frac{17}{6}\underset{\sim}{j} - \frac{1}{3}\underset{\sim}{k}$

MATCHED EXAMPLE 9

$3\underset{\sim}{i} - 13\underset{\sim}{j} - 12\underset{\sim}{k}$

MATCHED EXAMPLE 10

$-24\underset{\sim}{i} - 2\underset{\sim}{j} + 44\underset{\sim}{k}$

MATCHED EXAMPLE 11

a $\frac{1}{3}\left(-\underset{\sim}{i} + 2\underset{\sim}{j} + 2\underset{\sim}{k}\right)$

b $\frac{\sqrt{3}}{3}\left(\underset{\sim}{i} - 2\underset{\sim}{j} - 2\underset{\sim}{k}\right)$

MATCHED EXAMPLE 12

Since $\underset{\sim}{a} \cdot \underset{\sim}{b} = 0$, $\underset{\sim}{a}$ and $\underset{\sim}{b}$ are perpendicular.

MATCHED EXAMPLE 13

$M\left(\frac{17}{12}, -\frac{5}{3}\right)$

MATCHED EXAMPLE 14

Option E, $\underset{\sim}{a} \cdot \underset{\sim}{b} = 0$, is correct.

CHAPTER 2

MATCHED EXAMPLE 1

The locus is the straight line with the equation $2x - y + 2 = 0$.

MATCHED EXAMPLE 2

The equation of the ellipse could be

$\frac{(x-5)^2}{4} + \frac{(y-2)^2}{9} = 1$ or

$\frac{(x-1)^2}{4} + \frac{(y-2)^2}{9} = 1$.

MATCHED EXAMPLE 3

The parametric form is $\begin{cases} y = 6\sin(t) - 2 \\ x = 5\cos(t) + 1 \end{cases}$.

MATCHED EXAMPLE 4

$\frac{(y+1)^2}{16} - \frac{(x-2)^2}{49} = 1$

MATCHED EXAMPLE 5

$\frac{(x-4)^2}{36} + \frac{9y^2}{100} = 1$

MATCHED EXAMPLE 6

$r = \frac{7}{3 + 4\sin(\theta)}$

MATCHED EXAMPLE 7

a $x^2 + x + 2$ has no real roots, so there are no points of discontinuity.

b $g(x)$ is discontinuous at $x = 0$ or $x = 1$.

c $h(x)$ is discontinuous at $x = 3$ and $x = 2$.

MATCHED EXAMPLE 8

a $x = -2$ or $x = 3$

b $x = -\frac{2}{3}$ or $x = 3$

MATCHED EXAMPLE 9

The quotient is 4 and the remainder is $-14x + 12$.

MATCHED EXAMPLE 10

The quotient is $3x - 5$ and the remainder is $3x - 2$.

MATCHED EXAMPLE 11

a $\dfrac{5x-4}{x^2-x-2}=\dfrac{3}{x+1}+\dfrac{2}{x-5}$

b $\dfrac{4x^2+5x+8}{(x^2+5)(x+2)}=\dfrac{22x+1}{9(x^2+5)}+\dfrac{14}{9(x+2)}$

MATCHED EXAMPLE 12

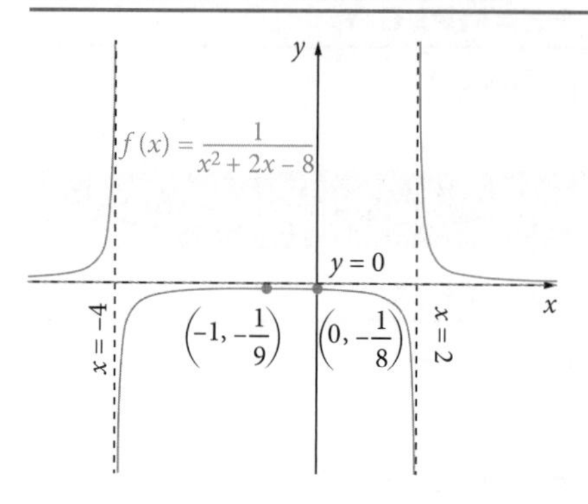

MATCHED EXAMPLE 13

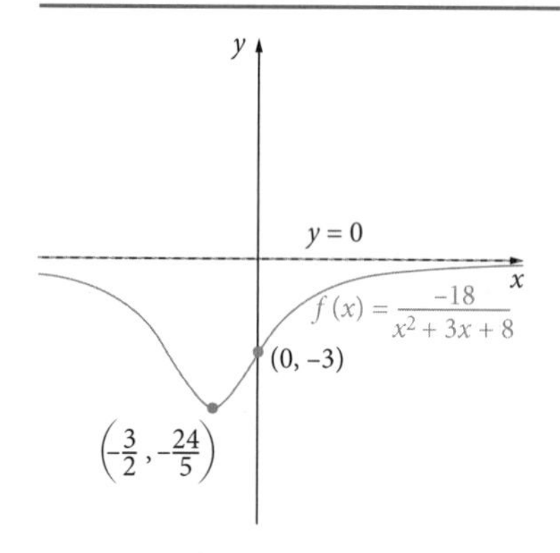

MATCHED EXAMPLE 14

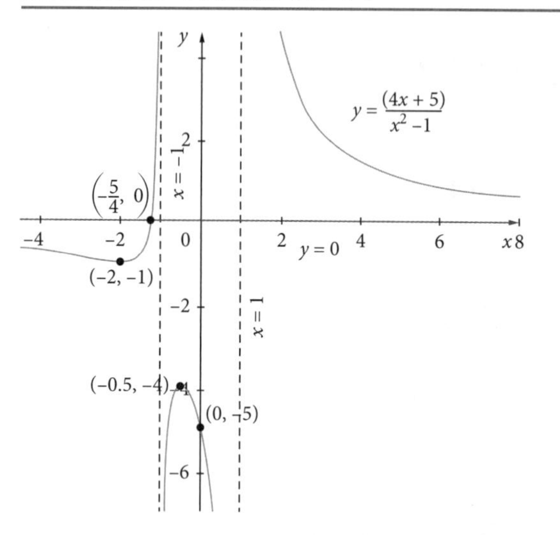

MATCHED EXAMPLE 15

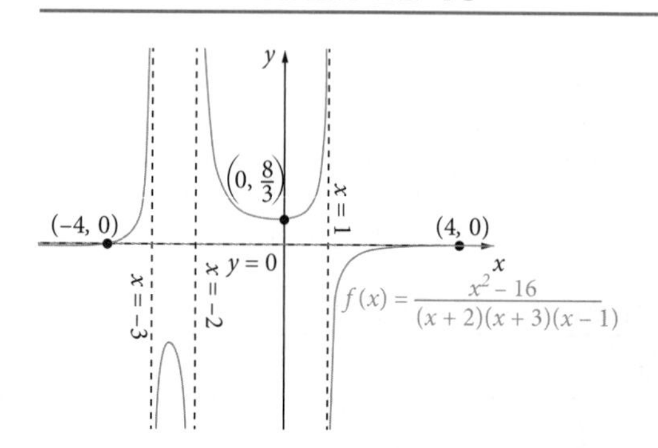

MATCHED EXAMPLE 16

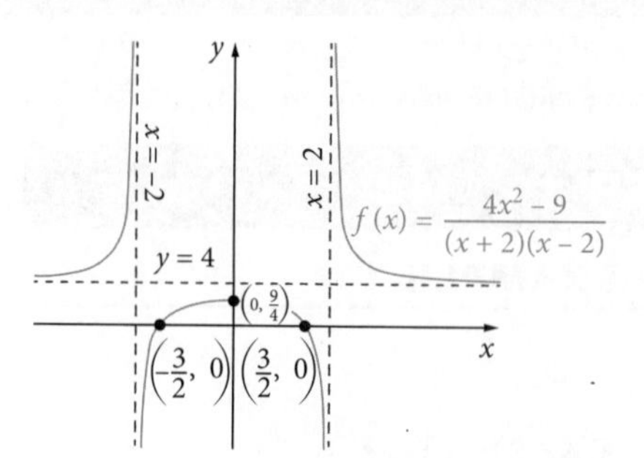

MATCHED EXAMPLE 17

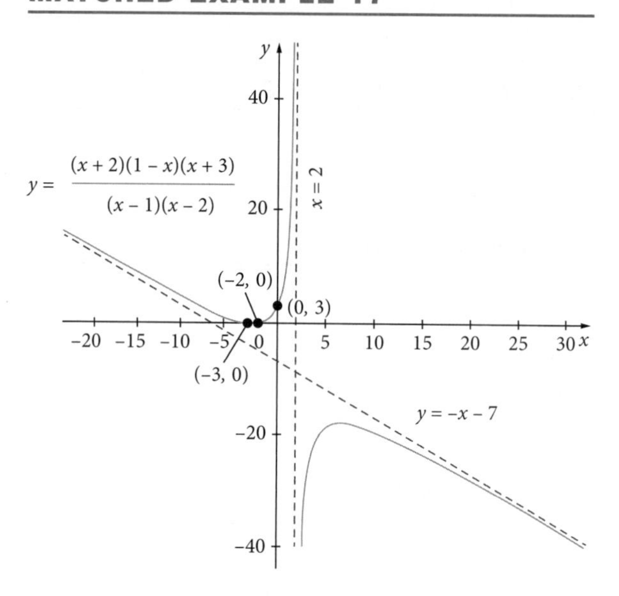

MATCHED EXAMPLE 18

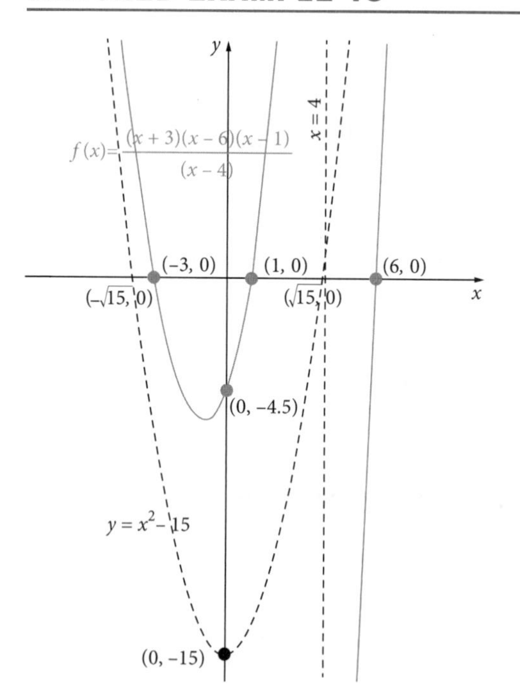

MATCHED EXAMPLE 19

a $x=-2$ or $x=\dfrac{10}{3}$

b $x \geq 5$ or $x \leq 1$

c $x \in (-4, 6)$

MATCHED EXAMPLE 20

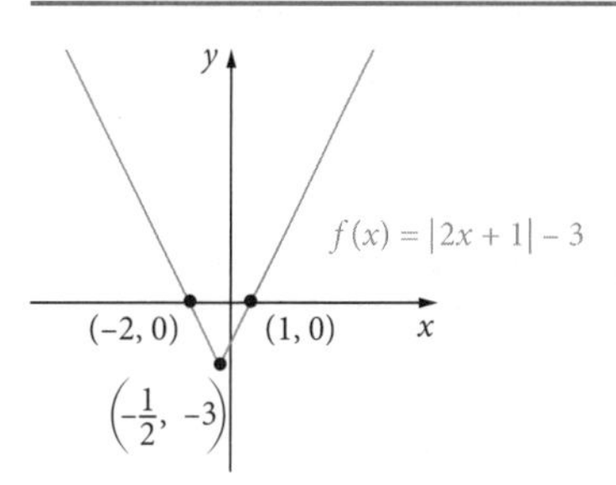

MATCHED EXAMPLE 21

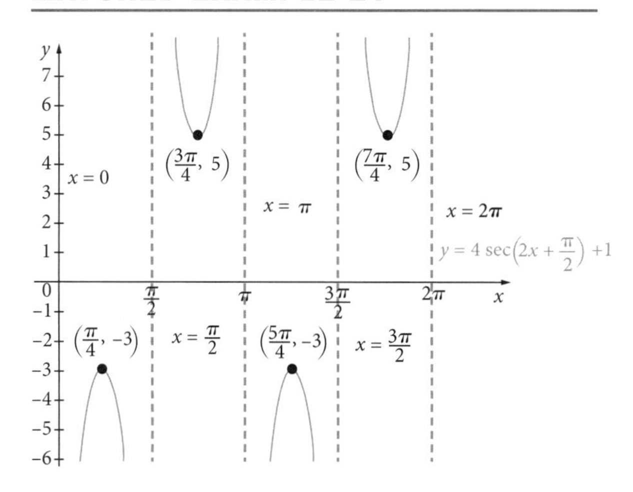

MATCHED EXAMPLE 22

$x = \frac{3\pi}{8} \pm n\pi$ for $n \in Z$

MATCHED EXAMPLE 24

$x = \frac{\pi}{6} + n\pi, x = \frac{\pi}{3} + n\pi$ or $x = \frac{\pi}{4} + n\pi$

MATCHED EXAMPLE 25

$\cot(x) = \frac{1}{\sqrt{3}}$

MATCHED EXAMPLE 26

$x = 3$ or $x = -2$

MATCHED EXAMPLE 27

a $\frac{3\pi}{2}$

b $\frac{\pi}{4}$

MATCHED EXAMPLE 28

The domain is $x \in [-12, -4]$ and the range is $y \in \left[-\frac{\pi}{2}, \frac{3\pi}{2}\right]$.

MATCHED EXAMPLE 29

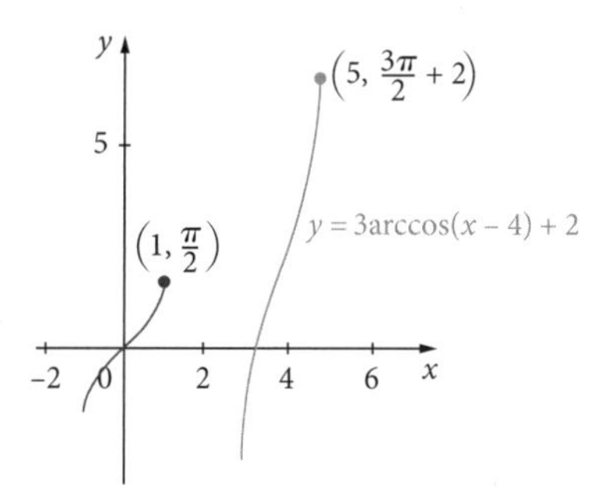

MATCHED EXAMPLE 30

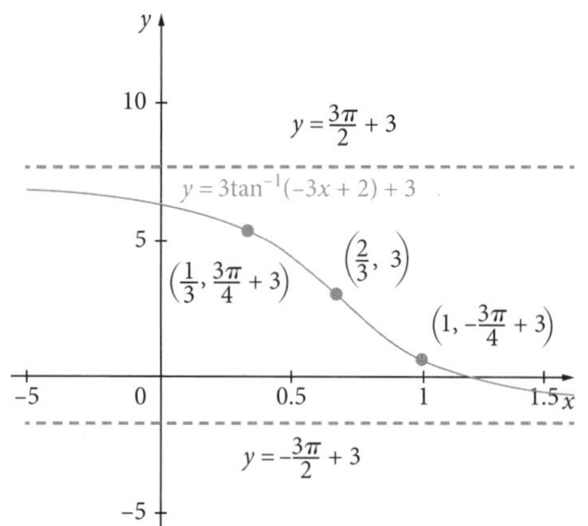

MATCHED EXAMPLE 31

a Domain $= \{x \in R: \frac{c-1}{b} \le x \le \frac{1+c}{b}, b > 0\}$

range $= \{y \in R: a \le y \le a + d\pi^2\}$

b Domain: $x > \frac{d}{c}$, range $b < y < \sqrt{\frac{a\pi}{2}} + b$

CHAPTER 3

MATCHED EXAMPLE 1

a 1 is a question, so it is not a statement.

Each of 2, 3, 4 and 5 is either true or false, so it is a statement.

b There are no compound statements. All the statements are atomic.

c 2 is a premise and 4 is an inference.

3 is an inference. Chen concludes that Lucy won because her strategic thinking skills are good.

4 is an inference. Lucy concludes that she played chess every day last year and therefore won.

d Each premise and its inference are true. The general conclusion is that playing chess every day helps to develop strategic thinking skills. Hence, the conversation is a logical argument.

MATCHED EXAMPLE 2

a **1** A specific conclusion is made from a general case.

2 This is deductive reasoning.

b **1** A general conclusion is made from a particular case (the observations made on every Friday in the last month).

2 This is inductive reasoning.

c **1** A specific conclusion is made, but it does not follow from the statements because it is not mentioned if it's peanuts.

2 This is not inductive or deductive reasoning.

MATCHED EXAMPLE 3

a **i** Let $x = 2$ and $y = 4$ be the two integers, so $x^2 + y^2 = 4 + 16 = 20$, an even number.

ii Let $x = 2$ and $y = 3$ be the two integers, so $x^2 + y^2 = 4 + 9 = 13$, an odd number.

b **i** Let $a = 4$ and $b = 2$

$\frac{a}{b}=\frac{4}{2}=2$ and $\frac{b}{a}=\frac{2}{4}=\frac{1}{2}$

So $\frac{a}{b}>\frac{b}{a}$.

ii Let $a = 2$ and $b = 4$

$\frac{a}{b}=\frac{2}{4}=\frac{1}{2}$ and $\frac{b}{a}=\frac{4}{2}=2$

So $\frac{a}{b}<\frac{b}{a}$.

c **i** When $n = 9$ is divided by 3, the remainder is 0.

ii When $n = 2$ is divided by 3, the remainder is not 0.

MATCHED EXAMPLE 4

a $t(n)=\frac{6n+2}{3}$, $n = 1, 2, 3, \ldots$

b $t(4)=\frac{26}{3}$, $t(5)=\frac{32}{3}$, $t(6)=\frac{38}{3}$

c $\frac{6n+2}{3}=2n+\frac{2}{3}$, $n = 1, 2, 3, \ldots$

Division by 3 produces even whole numbers and $\frac{2}{3}$. The divisor is 3 and the fraction remainder is $\frac{2}{3}$.

MATCHED EXAMPLE 5

a **If** dogs are good pets, **then** they are trained.

b **If** dogs are not loyal, **then** they are **not** good pets.

c **If** dogs are loyal and trained, **then** I like them.

d **If** dogs are **not** loyal, **then** they are **not** trained or they are **not** good pets.

e I like dogs **if and only if** they are trained.

MATCHED EXAMPLE 6

a $A \to B \wedge C$

b $A \to B \vee C$

c $(A \wedge \neg B) \to (\neg C \wedge D)$

MATCHED EXAMPLE 7

a An angle is a right angle if and only if it has a measure of 90°.

b Anita goes to the library if and only if it is a weekday.

MATCHED EXAMPLE 8

Proof.

Show that for $n \in N$, n is divisible by both 2 and 3 if and only if n is divisible 6.

1 Choose statements. X (necessary condition for Y): number divisible by both 2 and 3.

Y (sufficient condition for X): divisible by 6.

2 Show $X \to Y$. n is divisible by both 2 and 3 means $n=2\times3\times m=6m,\ \ m\in N$.

Hence n is divisible by 6.

3 Show $Y \to X$. n is divisible by 6 means $n=2\times3\times m,\ \ m\in N$.

Hence n is divisible by both 2 and 3.

MATCHED EXAMPLE 9

Proof

Show that for $n \in N$, $n^2 + 2$ odd $\leftrightarrow n$ is odd.

1 Choose statements. $X: n^2 +2$ is odd. $Y: n$ is odd.

2 Show $X \to Y$. $n^2 +2$ odd means $(n \times n) + 2$ is odd.

An odd number plus 2 is still odd.

Hence n is odd, since if n was even, the product of two even numbers plus 2 would be even.

3 Show $Y \to X$. n is odd, so the product of two odd numbers is odd.

An odd number plus 2 is still odd.

Therefore, $n^2 +2$ is odd.

MATCHED EXAMPLE 10

Proofs

a **1** Write the statements to be used. A: {3, 6, 9, 12, 15}, B: {2, 4, 6, 8, 10, 12, 14}

2 Find the set of numbers that satisfy $\neg(A \vee B)$. $A \vee B$: {6, 12} Numbers common to A and B.

$\neg(A \vee B)$: {1, 2, 3, 4, 5, 7, 8, 9, 10, 11, 13, 14, 15} Numbers in the given set that do not include {6, 12}.

3 Find the set of numbers that satisfy $\neg A \wedge \neg B$. $\neg A$: {1, 2, 4, 5, 7, 8, 10, 11, 13, 14} Numbers in the given set that are not in A. $\neg B$: {1, 3, 5, 7, 9, 11, 13, 15} Numbers in the given set that are not in B.

$\neg A \wedge \neg B$: {1, 2, 3, 4, 5, 7, 8, 9, 10, 11, 13, 14, 15} Numbers in the given set that are not in A and B.

b **1** Write the statements to use. A: He sat on a wall. B: He had a great fall.

2 Express $\neg(A \wedge B)$ with words. **Not** (He sat on a wall **and** he had a great fall).

3 Express $\neg(A \wedge B)$ with correct grammar. He did not sit on a wall or have a great fall.

4 Express $\neg A \vee \neg B$ in word form. **not** He sat on a wall **or not** He had a great fall.

5 Express $\neg A \vee \neg B$ using correct grammar. He did not sit on a wall or have a great fall.

c 1 Write in the form $\neg(X \vee Y) \neg(A \vee B \vee C) = \neg((A \vee B) \vee C)$

2 Apply De Morgan's law twice. $\neg(A \vee B \vee C) = \neg(A \vee B) \wedge \neg C = \neg A \wedge \neg B \wedge \neg C$

MATCHED EXAMPLE 11

Proofs

a 1 State what is to be proved. P: even number. Q: $n^2 + 2n + 4$ is divisible by 4.

Show $P \rightarrow Q$.

2 Write the first statement using known information. n is an even number, so it is divisible by 2.

$n = 2p$

Squaring on both sides

$n^2 = 4p^2$ which means n^2 is divisible by 4.

We can multiply by 2 to $n = 2p$, to get

$2n = 4p$ which means $2n$ is divisible by 4.

And 4 is divisible by 4.

State the sum.

$n^2 + 2n + 4 = 4p^2 + 4p + 4$

$= 4(p^2 + p + 1)$

$= 4k$, integer k.

3 Establish the conclusion. $4k$ is divisible by 4.

Hence $n^2 + 2n + 4$ is divisible by 4.

b 1 Write the statements to use. A: n is an integer.

B: $8^n - 1$ is divisible by 7.bn

2 Write the first statement using known information.

$8^n = 8^{n+1-1} = 8 \times 8^{n-1}$

This means 8^n is divisible by 8.

We know $a^n - b^n$ is divisible by $a - b$.

This means $8^n - 1^n$ is divisible by $8 - 1$.

Or $8^n - 1$ is divisible by 7.

3 Establish the conclusion. For any integer n, $8^n - 1$ is divisible by 7.

MATCHED EXAMPLE 12

a 'For all integers, x, there exists an integer y such that $y = x^2 + 1$'.

b 'For all integers, p and q, where q *is* not equal to zero, there exists a y in the set of real numbers such that $y = \frac{p}{q}$'.

MATCHED EXAMPLE 13

a $\forall x\, \exists y \left(y = \frac{1}{x} \right)$

b $\forall b\, \exists a\, (a = \cos(b))$

MATCHED EXAMPLE 14

a The statement is false.

b The statement is true.

MATCHED EXAMPLE 15

Proof

1 Translate the statement to omit quantifiers. For all rational values x, y, $x \cdot y$ is rational.

2 Write as a conditional statement, $P \rightarrow Q$. P: x and y are rational numbers.

Q: $x \cdot y$ is rational.

3 Use P to deduce Q. Let $x = \frac{a}{b}$, $y = \frac{c}{d}$, $a, b, c \neq 0, d \neq 0 \in Z$

$x \cdot y = \frac{a}{b} \cdot \frac{c}{d} = \frac{ac}{bd} = \frac{l}{m}$, where $l, m \neq 0 \in Z$

4 State the proof. $x \cdot y$ is the ratio of two integers, hence the product of two rational numbers is rational.

MATCHED EXAMPLE 16

Proofs

a 1 Write the statement using quantifiers. $\exists n \in N(n^2 + 7n + 12)$ is prime.

2 Write the negation. $\forall n \in N(n^2 + 7n + 12)$ is not prime.

3 Show that the statement is true. $n^2 + 7n + 12 = (n + 3)(n + 4)$

$n + 3 > 1$ and $n + 4 > 1$

So $(n + 3)(n + 4)$ is the product of two integers, each of which is greater than 1.

Hence $n^2 + 7n + 12$ is not prime.

b 1 Write the statement using quantifies. $\forall n \in Z \exists k \in Z(n = 2k)$

2 Write the negation. $\exists n \in Z \forall k \in Z(n \neq 2k)$

3 Show the statement to be true. Provide a counterexample.

$n = 3$ cannot be written as $2k$.

MATCHED EXAMPLE 18

Proof

1 Express the statement $P \rightarrow Q$ in the form $\neg Q \rightarrow \neg P$.

P: $n^2 - 1$ is not divisible by 8.

Q: n is not an odd positive integer.

$P \rightarrow Q$ if $n^2 - 1$ is not divisible by 8, then n is not an odd positive integer.

2 State $\neg Q \rightarrow \neg P$ $\neg Q \rightarrow \neg P$ if n is an odd positive integer, then $n^2 - 1$ is divisible by 8.

3 Prove $\neg Q \rightarrow \neg P$ If n is an odd positive integer, it can be written in the form $4p + 1$ or $4p + 3$ for some integer p.

If $n = 4p + 1$,

$n^2 - 1 = (4p + 1)^2 - 1$

$= 16p^2 + 8p$

$= 8(2p^2 + p)$, which is divisible by 8.

If $n = 4p + 3$,

$n^2 - 1 = (4p + 3)^2 - 1$

$= 16p^2 + 24p + 8$

$= 8(2p^2 + 3p + 1)$, which is divisible by 8.

Hence, $\neg Q \rightarrow \neg P$ is true, so $P \rightarrow Q$ is true.

MATCHED EXAMPLE 19

Proof

1 Assume the statement is false and write it as an equation. Assume that $\sqrt{5}$ is rational and can be written as $\sqrt{5} = \frac{a}{b}$, where $a, b \in Z$ have no common factors other than 1.

2 Rearrange the equation and form conclusions about the types of variables used. $a^2 = 5b^2$

So 5 *divides* a^2. Hence 5 *divides* a.

3 Write the equation in terms of the types of variables and rearrange. So let $a = 5r$, for some $r \in Z$.

$(5r)^2 = 5b^2$

$5r^2 = b^2$. So 5 divides b^2. Hence 5 divides b.

4 State the contradiction. Therefore, 5 is a common factor of a and b. This is a contradiction.

5 State the conclusion. The assumption that $\sqrt{5}$ is rational leads to a contradiction, so the original assumption that $\sqrt{5}$ is rational is false.

MATCHED EXAMPLE 20

1 Assume the statement is false.

Assume a^2 is even then a is odd.

2 Rewrite the expression according to the types of numbers used.

a is odd, so $a = 2n + 1$, n is an integer.

3 Expand and rearrange the expression and form a conclusion.

$a^2 = (2n + 1)^2$

$a^2 = 4n^2 + 4n + 1$

$= 2(2n^2 + 2n) + 1$

This shows a^2 is odd.

4 State what must be true to avoid a contradiction.

Hence $a^2 = 2k + 1$.

5 Explain the contradiction, testing each case.

This is not possible because a^2 is even but $2k + 1$ is odd.

6 State the conclusion.

The assumption that a is odd is wrong, so a is even.

MATCHED EXAMPLE 21

Proof

1 Prove the base step. Let $P(n) = n(2n + 1)$.

True for $n = 1$.

$3 = 1(2 \times 1 + 1)$

2 State the hypothesis and the required expression. Assume the conjecture is true for $n = k$, $(k > 1)$.

For $n = k + 1$, the sum is the kth sum + the $(k + 1)$th term.

The $(k + 1)$th term is $4(k + 1) - 1 = 4k + 3$

$P(k + 1) = (k)(2k + 1) + 4k + 3$

3 Write the function in the required form.

$= 2k^2 + k + 4k + 3$

$= 2k^2 + 5k + 3$

$= (k + 1)(2k + 3)$

$= (k + 1)(2(k + 1) + 1)$

$= P(k + 1)$

4 State the conclusion. The hypothesis is true for $n = k + 1$, but because k is an arbitrary value, the hypothesis must be true for all values of n.

MATCHED EXAMPLE 22

Proof

1 Prove the base step. Let $P(n) = \frac{n}{6}(n+1)(2n+1)$.

True for $n = 1$.

$1 = \frac{1}{6}(1+1)(2+1)$.

2 State the hypothesis and the required expression. Assume the conjecture is true for $n = k$, $(k > 1)$.

For $n = k + 1$, the sum is the kth sum + the $(k + 1)$th term.

The $(k + 1)$th term is $(k + 1)^2$.

$P(k+1) = \frac{k}{6}(k+1)(2k+1) + (k+1)^2$

3 Write the function in the required form.

$= \frac{k+1}{6}[k(2k+1) + 6(k+1)]$

$= \frac{k+1}{6}[2k^2 + 7k + 6]$

$= \frac{(k+1)(k+2)(2k+3)}{6}$

$= \frac{(k+1)(k+2)[2(k+1)+1]}{6}$

4 State the conclusion. The hypothesis is true for $n = k + 1$, but k is an arbitrary value, so the hypothesis is true for all values of n.

MATCHED EXAMPLE 23

Proof

1 Prove the base step. $P(n) = 3^{2n} - 1$.

True for $n = 1$. $3^2 - 1 = 8$ is divisible by 8.

2 State the hypothesis and the required expression. Assume the conjecture is true for $n = k$, $(k > 1)$.

That is, that $3^{2k} - 1$ is divisible by 8.

$P(k + 1) = 32^{(k+1)} - 1$

3 Write the function as the multiple of 8.

$= 3^{2k} \cdot 3^2 - 1$

$= 3^{2k} \cdot 9 - 9 + 8$

$= (3^{2k} - 1)9 + 8$

$= (8m)9 + 8$

$= 8(9m + 1)$

$P(k+1)$ is divisible by 8.

4 State the conclusion. The hypothesis is true for $n = k + 1$, but k is an arbitrary value, so the hypothesis is true for all values of n.

MATCHED EXAMPLE 24

Proof

1 Prove the base step. $P(n)=\dfrac{\sin(nx)}{2\sin\left(\dfrac{x}{2}\right)}$

True for $n = 1$. LHS $= \cos\left(\dfrac{x}{2}\right)$.

$$\text{RHS}=\frac{\sin(x)}{2\sin\left(\frac{x}{2}\right)}=\frac{2\sin\left(\frac{x}{2}\right)\cos\left(\frac{x}{2}\right)}{2\sin\left(\frac{x}{2}\right)}=\cos\left(\frac{x}{2}\right).$$

2 State the hypothesis and the required expression. Assume the conjecture is true for $n = k$, $(k > 1)$

$$P(k+1)=\frac{\sin(kx)}{2\sin\left(\frac{x}{2}\right)}+\cos\left(kx+\frac{x}{2}\right).$$

3 Write the function in the required form.

$$=\frac{\sin(kx)+2\cos\left(kx+\frac{x}{2}\right)\sin\left(\frac{x}{2}\right)}{2\sin\left(\frac{x}{2}\right)}$$

$$=\frac{\sin(kx)+2\sin\left(\frac{x}{2}\right)\left[\cos(kx)\cos\left(\frac{x}{2}\right)-\sin(kx)\sin\left(\frac{x}{2}\right)\right]}{2\sin\left(\frac{x}{2}\right)},$$

using the identity $\cos(A+B)=\cos A\cos B-\sin A\sin B$

$$=\frac{\sin(kx)+2\sin\left(\frac{x}{2}\right)\cos(kx)\cos\left(\frac{x}{2}\right)-2\sin\left(\frac{x}{2}\right)\sin(kx)\sin\left(\frac{x}{2}\right)}{2\sin\left(\frac{x}{2}\right)}$$

$$=\frac{\sin(kx)+\sin(x)\cos(kx)-2\sin^2\left(\frac{x}{2}\right)\sin(kx)}{2\sin\left(\frac{x}{2}\right)}$$

$$=\frac{\sin(x)\cos(kx)+\left[1-2\sin^2\left(\frac{x}{2}\right)\right]\sin(kx)}{2\sin\left(\frac{x}{2}\right)}$$

$$=\frac{\sin(x)\cos(kx)+\cos(x)\sin(kx)}{2\sin\left(\frac{x}{2}\right)}$$

$$=\frac{\sin((k+1)x)}{2\sin\left(\frac{x}{2}\right)}$$

$= P(k+1)$

4 State the conclusion. The hypothesis is true for $n = k + 1$, but k is an arbitrary value, so the hypothesis is true for all values of n.

CHAPTER 4

MATCHED EXAMPLE 1

a

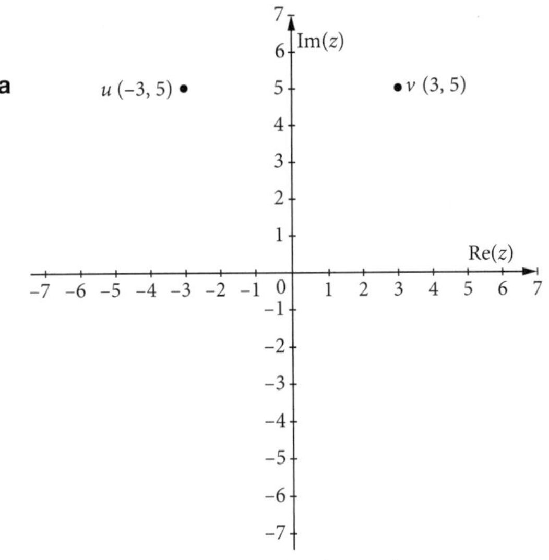

b They are reflections of each other in the y-axis.

c

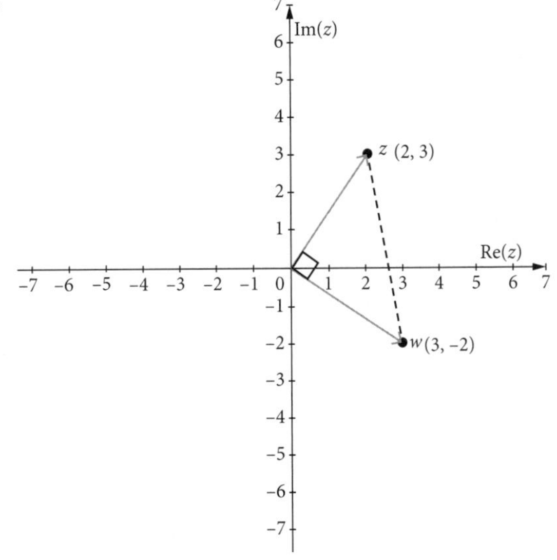

d They are perpendicular to each other.

MATCHED EXAMPLE 2

a $= 3 - 8i$

b $-4 + 9i$

c $-3 - 11i$

d $\dfrac{13}{26}-\dfrac{13}{26}i$ or $0.5 - 0.5i$

MATCHED EXAMPLE 3

a $v=4\,\text{cis}\left(\dfrac{\pi}{6}\right), z=2\sqrt{2}\,\text{cis}\left(\dfrac{\pi}{4}\right)$

b $u=2\sqrt{2}+2\sqrt{2}i,\ w=\sqrt{3}-i$

c 1 $uv=16\,\text{cis}\left(\dfrac{5\pi}{12}\right)$

2 $wv=8\,\text{cis}(0)$

d **1** $\frac{v}{u} = \text{cis}\left(-\frac{\pi}{12}\right)$

2 $\frac{z}{u} = \frac{1}{\sqrt{2}}\text{cis}(0)$

MATCHED EXAMPLE 4

a The equation is $|z - 3 + 2i| = k|-2 + 4i|$ for $k \in R$

b The cartesian equation is $2x + y - 4 = 0$

MATCHED EXAMPLE 5

a The equation is $|z - 2 - 3i| = 4$

b

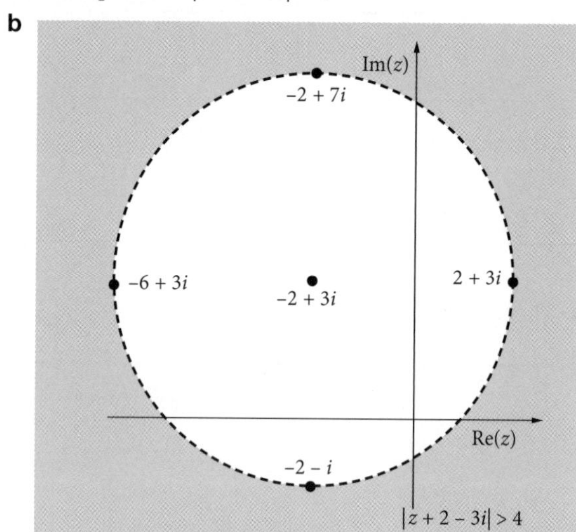

MATCHED EXAMPLE 6

1 It is an ellipse with foci $-1 - 4i$, $2 + 5i$ and major semi-axis length 5. The minor semi-axis length is given by $b^2 = 25 - 0.25 \times 90 = 2.5$, so the minor semi-axis length is 1.58.

2

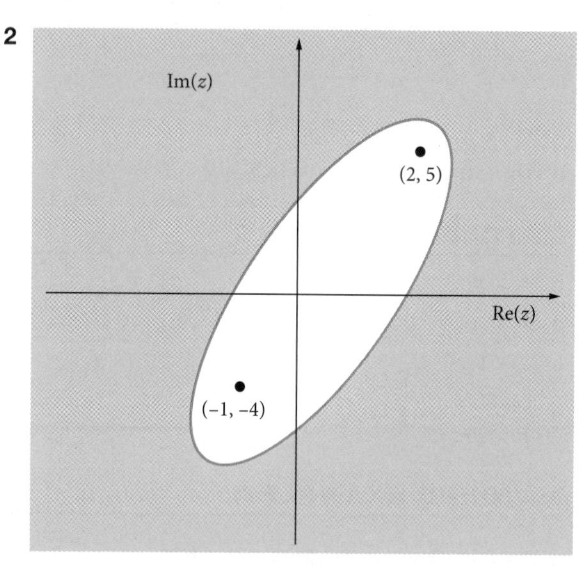

MATCHED EXAMPLE 7

1 It is a semicircle on the diameter $3i - 2$ to $5 + i$, including the interior and the circumference, but not the diameter.

2

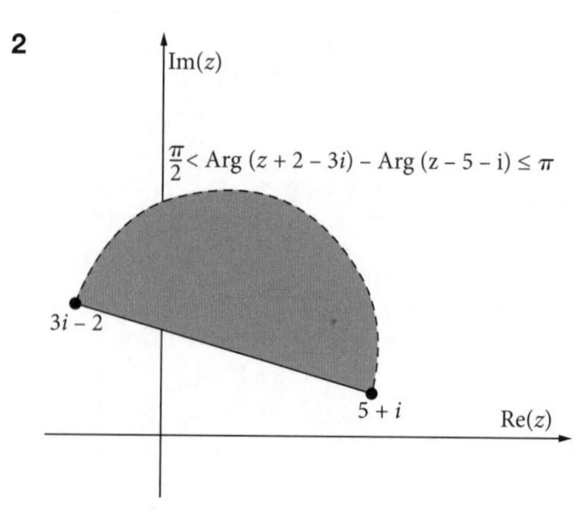

MATCHED EXAMPLE 8

a It is the intersection of two regions.

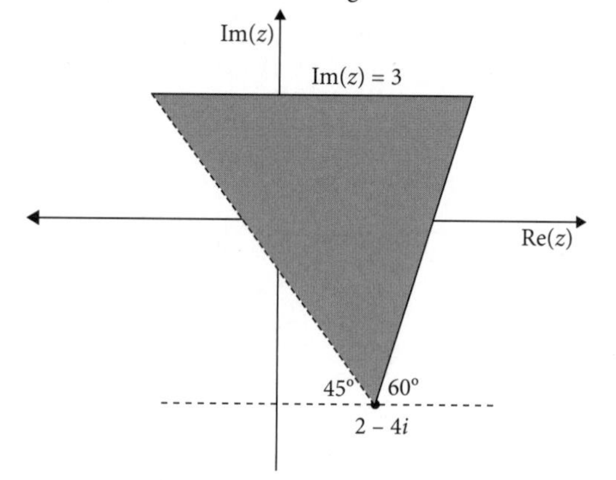

b $\frac{49(3+\sqrt{3})}{6}$ square units

MATCHED EXAMPLE 9

$z = 2\sqrt{6}\,\text{cis}\left(-\frac{\pi}{3}\right)$

$z^4 = -288 + 288\sqrt{3}i$

MATCHED EXAMPLE 10

$z = 1, \text{cis}\left(\pm\frac{2\pi}{5}\right), \text{cis}\left(\pm\frac{4\pi}{5}\right),$

or $z = \text{cis}\left(\frac{2k\pi}{5}\right)$ for $k = -2, -1, 0, 1, 2$.

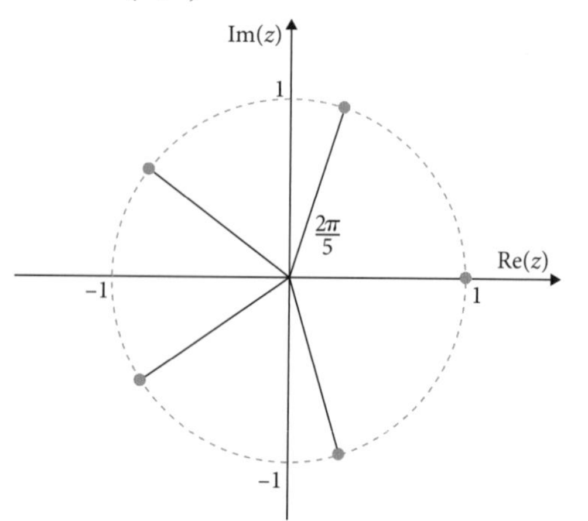

MATCHED EXAMPLE 11

$w = 4\text{ cis}\left(\frac{-9\pi}{12}\right) = 4\text{ cis}\left(\frac{-3\pi}{4}\right)$,

$w = 4\text{ cis}\left(\frac{-\pi}{12}\right)$, or $w = 4\text{ cis}\left(\frac{7\pi}{12}\right)$

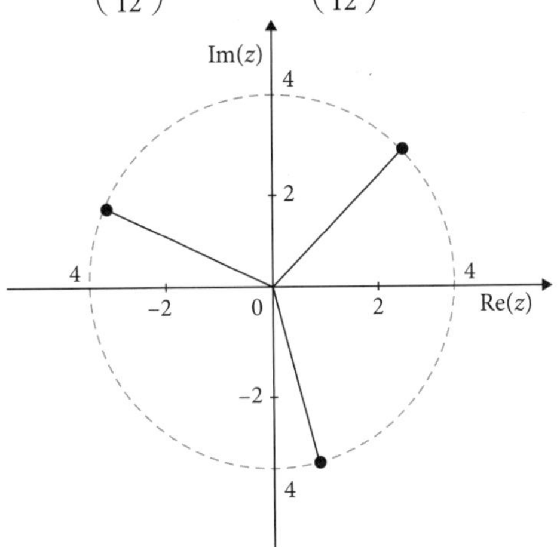

MATCHED EXAMPLE 12

a $(z-2-\sqrt{2}\,i)(z-2+\sqrt{2}\,i)$

b $\left(z+\frac{2}{3}-\frac{\sqrt{2}}{3}i\right)\left(z+\frac{2}{3}+\frac{\sqrt{2}}{3}i\right)$

MATCHED EXAMPLE 13

The remainder is $8 - 20i$.

MATCHED EXAMPLE 14

$P(5+i) = (5+i)^3 - (5+3i)(5+i)^2 - (11-16i)(5+i) + 51 - 21i$

$= 125 + 75i + 15i^2 + i^3 - (120 + 72i + 50i + 30i^2) - (55 + 11i - 80i - 16i^2) + 51 - 21i$

$= 125 + 75i - 15 - i - (120 + 122i - 30) - (55 - 69i + 16) + 51 - 21i$

$= 110 + 74i - 120 - 122i + 30 - 55 + 69i - 16 + 51 - 21i$

$= 0$

$P(3) = 3^3 - (5+3i) \times 3^2 - (11-16i) \times 3 + 51 - 21i$

$= 27 - 45 - 27i - 33 + 48i + 51 - 21i$

$= 0$

Since $P(5+i) = 0$, $z - (5+i)$ is a factor of $P(z)$.

Since $P(3) = 0$, $z - 3$ is a factor of $P(z)$.

The remaining factor of $P(z)$ is $z + 3 - 2i$.

MATCHED EXAMPLE 15

a $z^2 + 4i = \left(z-\sqrt{2}+\sqrt{2}i\right)\left(z+\sqrt{2}-\sqrt{2}i\right)$

b $z^2 + (4-2i)z + 3 - 2i = (z+1)(z+3-2i)$

MATCHED EXAMPLE 16

a $z^3 + z^2 i - z^2 - 4z + 4 - 4i = (z-2)(z+2)(z-1+i)$

b $z^3 - 2z^2 + (10-3i)z - 9 + 3i = (z-1)(z+3i)(z+1+3i)$

MATCHED EXAMPLE 17

a The solutions are $\pm 9i$.

b The solutions are $2 \pm i$.

c The solutions are $\frac{-7 \pm 3\sqrt{7}i}{4}$.

MATCHED EXAMPLE 18

$4x^2 - 4x + 13 = 0$

MATCHED EXAMPLE 19

a $-\frac{\sqrt{14}}{2}-\frac{\sqrt{14}}{2}i,\ \frac{\sqrt{14}}{2}+\frac{\sqrt{14}}{2}i$

b $z = 3i,\ -\frac{i}{2}$

c $-\frac{\sqrt{6}}{2}+\frac{3\sqrt{2}-2\sqrt{3}}{2}i,\ \frac{\sqrt{6}}{2}-\frac{3\sqrt{2}-2\sqrt{3}}{2}i$

MATCHED EXAMPLE 20

$7 - 5i$

MATCHED EXAMPLE 21

$z = 1 - 3i$, $z = 1 + 3i$ or $z = \frac{3}{5}$

MATCHED EXAMPLE 22

The solutions are $z = -2, 3, \frac{1}{2}+\frac{\sqrt{5}}{2}i, \frac{1}{2}-\frac{\sqrt{5}}{2}i$.

MATCHED EXAMPLE 23

$z = -1$, $z = 3$ or $z = -2 + 5i$

MATCHED EXAMPLE 24

The solutions are $z = 2i$, $z = -2i$, $z = \frac{3}{2}i$, $z = -\frac{3}{5}i$.

CHAPTER 5

MATCHED EXAMPLE 1

$\frac{dy}{dx} = 4x^2(5x-9)$

MATCHED EXAMPLE 2

$\frac{dy}{dx} = \frac{-2(3x^2+6x+4)}{x^3(x+2)^2}$

MATCHED EXAMPLE 3

$\frac{dy}{dx} = 5x^4(4x^3+5)^4(16x^3+5)$

MATCHED EXAMPLE 4

$\frac{dy}{dx} = -\frac{7}{18}$

MATCHED EXAMPLE 5

$\frac{dy}{dx} = -6\sin(4x)\cos(2x)$

MATCHED EXAMPLE 6

$\therefore \frac{dy}{dx}=0$

MATCHED EXAMPLE 7

The gradient of the curve $f(x)=\frac{\sin^2(x)}{\cos(x)}$ at $x=\frac{\pi}{6}$ is $\frac{7}{6}$.

MATCHED EXAMPLE 8

$\therefore \frac{dy}{dx}=-\frac{1}{\sqrt{4-x^2}}, x\in(-2, 2)$

MATCHED EXAMPLE 9

$\frac{4}{\sqrt{7}}$

MATCHED EXAMPLE 10

$44e^8$

MATCHED EXAMPLE 11

$\frac{dy}{dx}=-3x+6x\log_e\left(\frac{1}{6x}\right)$

MATCHED EXAMPLE 12

a $\frac{d^2y}{dx^2}=120x^3-12\cos(x)$

b At $x=0$, $\frac{d^2y}{dx^2}=-12$.

MATCHED EXAMPLE 13

$\left(-\frac{5}{6},-\frac{1315}{54}\right)$ is a non-stationary point of inflection.

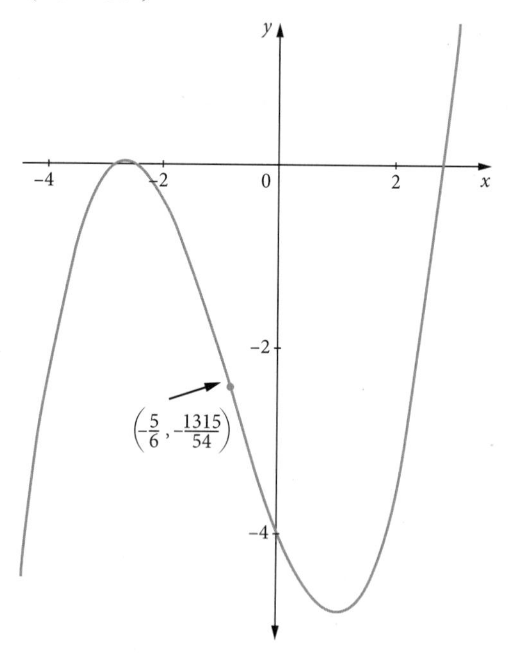

MATCHED EXAMPLE 14

$\left(\frac{2}{3},-\frac{52}{27}\right)$ is a non-stationary point of inflection.

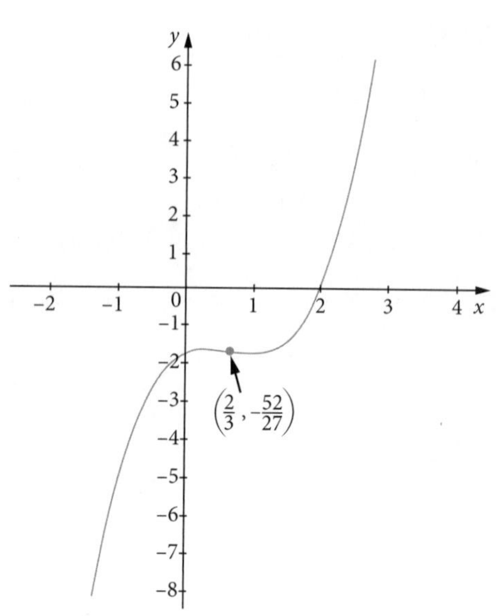

At $x=\frac{2}{3}$, the graph changes from concave down to concave up.

MATCHED EXAMPLE 15

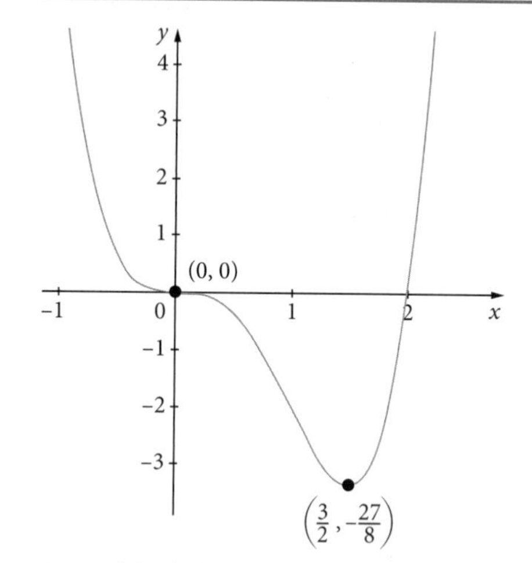

Around (0,0), the gradient shifts from negative to negative.

Around $\left(\frac{3}{2},-\frac{27}{8}\right)$ the gradient goes from negative to positive.

(0,0) is a stationary point of inflection

$\left(\frac{3}{2},-\frac{27}{8}\right)$ is a local minimum point.

MATCHED EXAMPLE 16

$\frac{dr}{dt}=\frac{1}{18}$ m/second

MATCHED EXAMPLE 17

$\frac{dh}{dt}=\frac{3}{10\pi}$ cm/min

MATCHED EXAMPLE 18

a $\frac{dy}{dx}=\frac{2x^2+y}{x-x^2}$

b $\frac{dy}{dx}=0$

9780170464079

CHAPTER 6

MATCHED EXAMPLE 1

a $y = \frac{(x-3)^2}{8} - 1$

The curve is a parabola that opens upwards with vertex (3, −1).

b $\left(\frac{x+2}{3}\right)^2 - \left(\frac{y-1}{4}\right)^2 = 1$

The curve is a horizontal hyperbola with centre (−2, 1).

MATCHED EXAMPLE 2

The shape is a sine curve of magnitude 2 along the line $z = -3x$ perpendicular to x–z plane.

MATCHED EXAMPLE 3

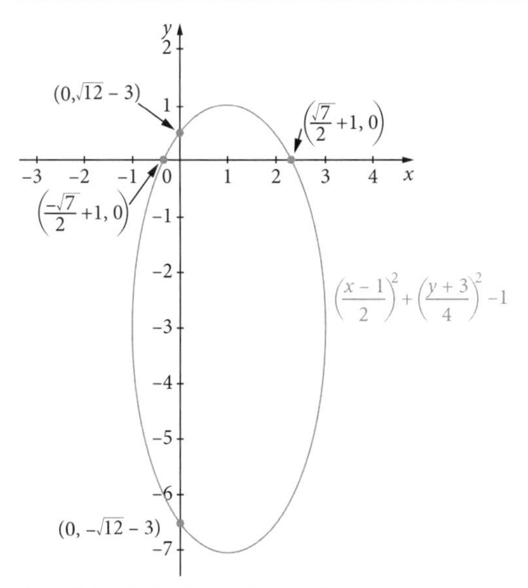

MATCHED EXAMPLE 4

a $\underset{\sim}{r}(t) = (2+t)\underset{\sim}{i} + (1-2t)\underset{\sim}{j}$ or $x = 2+t,\ y = 1-2t,\ t \in R$

b $\underset{\sim}{r}(t) = (1-2t)\underset{\sim}{i} + (-t)\underset{\sim}{j} + 2\underset{\sim}{k}$ or $x = 1-2t,\ y = -t,$
$z = 2\ t \in R$

MATCHED EXAMPLE 5

a $4x + 3y + 1 = 0$

b $\frac{x-2}{3} = \frac{y+1}{-2} = \frac{z}{-2}$

MATCHED EXAMPLE 6

a $\underset{\sim}{r}(t) = (2-t)\underset{\sim}{i} + (3-t)\underset{\sim}{j} + (-2+3t)\underset{\sim}{k},\ t \in R,\ 0 \le t \le 1$

b Only the point (1.5, 2.5, –0.5) is on the line segment PQ.

MATCHED EXAMPLE 7

$\underset{\sim}{r}(t) = (-2+12t)\underset{\sim}{i} + (1+4t)\underset{\sim}{j} + (5+3t)\underset{\sim}{k}$

MATCHED EXAMPLE 8

a (4, –5, 4) is not on Γ_1.

b (10, –10, 8) is on Γ_1.

MATCHED EXAMPLE 9

$-13\underset{\sim}{i} + 7\underset{\sim}{j} - 7\underset{\sim}{k}$ is a normal and $\frac{1}{\sqrt{267}}(-13\underset{\sim}{i} + 7\underset{\sim}{j} - 7\underset{\sim}{k})$ a unit normal to the plane Γ_1.

MATCHED EXAMPLE 10

$\underset{\sim}{n} = -5\underset{\sim}{i} + \underset{\sim}{j} - \underset{\sim}{k}$ is a normal to the plane Γ_1.

MATCHED EXAMPLE 11

a The equation of the plane is $3x - 2y + z = -7$.

b The vector equation is $(\underset{\sim}{i} + 3\underset{\sim}{j} - \underset{\sim}{k}) \cdot \underset{\sim}{p} = 4$.

MATCHED EXAMPLE 12

The equation of the plane containing the points (2, 1, –4), (–2, –2, 3) and (1, 4, 3) is $-14x + 7y - 5z = -41$.

MATCHED EXAMPLE 13

The equation of the plane is $-2x - y + 5z = -12$

MATCHED EXAMPLE 14

The plane is $9x + 15y - 8z = -17$.

MATCHED EXAMPLE 15

a (3, –2, 6) is not on the plane.

b (2, –3, –6) is on the plane.

c (1, 2, –4) is not on the plane.

MATCHED EXAMPLE 16

The planes $3x - 2y + 4z = 2$ and $x + 2y - 2z = 4$ intersect in the line $\underset{\sim}{r}(t) = 2t\underset{\sim}{i} + (-5t+5)\underset{\sim}{j} + (3-4t)\underset{\sim}{k}$.

CHAPTER 7

MATCHED EXAMPLE 1

$\frac{x^6}{2} + \frac{5}{4x^4} + 2x + c$

MATCHED EXAMPLE 2

$\int 2e^{5x} + \frac{7}{x}\,dx = \frac{2e^{5x}}{5} + 7\log_e|x|$

MATCHED EXAMPLE 3

$\int \frac{5}{7x-3}\,dx = \frac{5}{7}\log_e|7x-3| + c$

MATCHED EXAMPLE 4

$\frac{5(2x+1)^4}{8} + c$

MATCHED EXAMPLE 5

$-\frac{1}{3}\cos(3x) - \frac{1}{5}\sin(5x) + c$

MATCHED EXAMPLE 6

$\frac{1}{2}\log_e(3)$

MATCHED EXAMPLE 7

$\sin^{-1}\left(\frac{x}{7}\right)+c$

MATCHED EXAMPLE 8

$\int \frac{-2}{\sqrt{4-25x^2}}\,dx = \frac{2}{5}\cos^{-1}\left(\frac{5}{2}x\right)+c$

MATCHED EXAMPLE 9

$\frac{1}{9}\tan^{-1}\left(\frac{x}{9}\right)+c$

MATCHED EXAMPLE 10

$\frac{1}{20}\tan^{-1}\left(\frac{4x}{5}\right)+c$

MATCHED EXAMPLE 11

$\frac{1}{4}\tan^{-1}\left(\frac{x+3}{4}\right)+c$

MATCHED EXAMPLE 12

$\frac{2\pi}{3}$

MATCHED EXAMPLE 13

$\frac{\pi}{12\sqrt{3}}$

MATCHED EXAMPLE 14

$\frac{2(2x^3+x)^{\frac{3}{2}}}{3}+c$

MATCHED EXAMPLE 15

$\frac{3(x^3+4)^5}{5}+c$

MATCHED EXAMPLE 16

$-\frac{\cos^6(x)}{6}+c$

MATCHED EXAMPLE 17

$\frac{2(3x-5)^{\frac{5}{2}}}{45}+\frac{10(3x-5)^{\frac{3}{2}}}{27}+c$

MATCHED EXAMPLE 18

$\frac{1}{4}$

MATCHED EXAMPLE 19

$\frac{26}{3}$

MATCHED EXAMPLE 20

$-\cos(x)+\frac{2}{3}\cos^3(x)-\frac{\cos^5(x)}{5}+c$

MATCHED EXAMPLE 21

$\frac{5}{2}x+\frac{3}{4}\sin(2x)+c$

MATCHED EXAMPLE 22

$\frac{1}{2}x-\frac{1}{20}\sin(10x)+c$

MATCHED EXAMPLE 23

$\frac{1}{3}\tan(3x)+x+c$

MATCHED EXAMPLE 24

$\frac{1}{8}\cos(4x)-\frac{1}{2}\cos(2x)+c$

MATCHED EXAMPLE 25

$\frac{1}{384}$

MATCHED EXAMPLE 26

$-\log_e|x+2|+2\log_e|x-2|+c$

MATCHED EXAMPLE 27

$\log_e\left(\frac{625}{576}\right)$

MATCHED EXAMPLE 28

$\int \frac{x-5}{(x+1)^2}\,dx = \log_e|x+1|+\frac{6}{(x+1)}+c$

MATCHED EXAMPLE 29

$\frac{dy}{dx} = 2\sin x\cos x$

$\int \sin x\cos x\,dx = \frac{1}{2}\sin^2 x+c$

MATCHED EXAMPLE 30

$\int \frac{x-3}{\sqrt{4-x^2}}\,dx = -\sqrt{4-x^2}-3\sin^{-1}\left(\frac{x}{2}\right)+c$

MATCHED EXAMPLE 31

$\int xe^{2x}\,dx = \frac{1}{2}xe^{2x}-\frac{1}{4}e^{2x}+c$

MATCHED EXAMPLE 32

$\int x^2\sin x\,dx = -x^2\cos x+2x\sin x+2\cos x+c$

MATCHED EXAMPLE 33

$\int \arcsin(x)\,dx = x\arcsin(x)+\sqrt{1-x^2}+c$

MATCHED EXAMPLE 34

$\int e^x \sin(x)\,dx = \frac{1}{2}e^x \sin(x) - \frac{1}{2}e^x \cos(x) + c$

CHAPTER 8

MATCHED EXAMPLE 1

6π square units

MATCHED EXAMPLE 2

$\frac{4}{\pi}$ units^2

MATCHED EXAMPLE 3

$\frac{8}{3}$ units^2

MATCHED EXAMPLE 4

$\frac{\pi}{2} - 1$ square units

MATCHED EXAMPLE 5

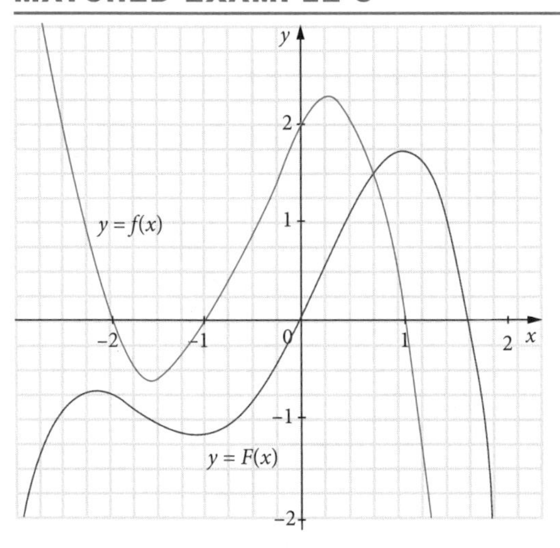

MATCHED EXAMPLE 6

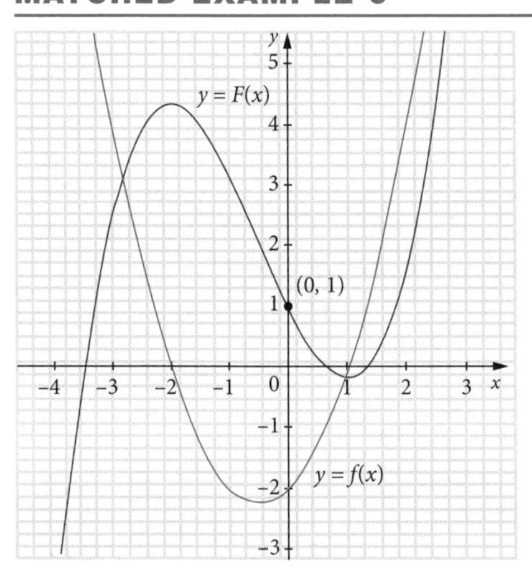

MATCHED EXAMPLE 7

$\frac{316\pi}{3}$ units^3

MATCHED EXAMPLE 8

π^2 units^3

MATCHED EXAMPLE 9

$\frac{64\pi}{15}$ units^3

MATCHED EXAMPLE 10

$\frac{2}{3}(\sqrt{27} - 1)$ units^2

MATCHED EXAMPLE 11

6 units

MATCHED EXAMPLE 12

8π units^2

MATCHED EXAMPLE 13

$\frac{\pi}{27}\left[145\sqrt{145} - 1\right]$ units^2

MATCHED EXAMPLE 14

$\sqrt{2}\pi$ units^2

CHAPTER 9

MATCHED EXAMPLE 1

a $\frac{dA}{dL} = kL$, where k is the constant of proportionality.

b $\frac{dx}{dt} = \frac{k}{x^2}$, where k is the constant of proportionality.

c $\frac{dP}{dt} = kP(1000 - P)$

where k is the constant of proportionality.

MATCHED EXAMPLE 3

$m = 2, m = 8$

MATCHED EXAMPLE 4

a $y = \tan^{-1}\left(\frac{x}{2}\right) + c$

b $y = \frac{t^4}{2} - 2t^2 + 3t + c$

c $p = 2\sin(y) + 3y + c$

MATCHED EXAMPLE 5

a $y = \frac{e^{2x}}{2} - 2x^2 + \frac{3}{2}$

b $y = \frac{2(3x-5)^{\frac{5}{2}}}{15} + \frac{1}{15}$

c $y = \frac{1}{2}x - \frac{1}{4}\sin(2x) + \frac{\pi}{4}$

ANSWERS

MATCHED EXAMPLE 6

$y = -\frac{1}{4}\cos(2x) + \sin(x) + c_1 x + c_2$

MATCHED EXAMPLE 7

The velocity of the particle is 160 m/s and the distance travelled by the particle is $457\frac{1}{3}$ m, after 10 s.

MATCHED EXAMPLE 8

a $y = \sqrt[3]{3x} + c$

b $x = t - 3 + c$

c $Q = \frac{1}{3}t^3$

MATCHED EXAMPLE 9

It will take 13 minutes for the oil's temperature to drop to 44°C.

MATCHED EXAMPLE 10

8936 million years

MATCHED EXAMPLE 11

a $\frac{dP}{dt} = kP$

b $P = 120e^{0.1t}$

c There will be 360 fish in 2023

MATCHED EXAMPLE 12

a $\frac{dM}{dt} = \frac{50 - M}{50}$

b $M = 50 - 45e^{-\frac{1}{50}t}$

c 19.8 kg

d

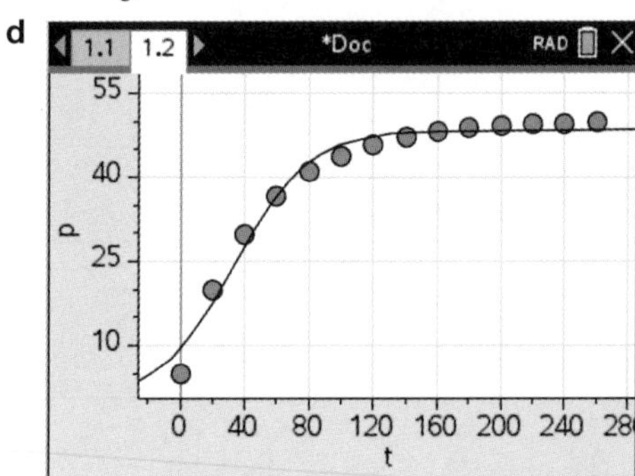

$\lim_{t \to \infty}\left(50 - 45e^{-\frac{1}{50}t}\right) = 50$

MATCHED EXAMPLE 13

a $\frac{dP}{dt} = \frac{kP(1500 - P)}{1500}$

b $P = \frac{1500}{1 + 5e^{-0.354t}}$

c $t = 4.92$ years

d

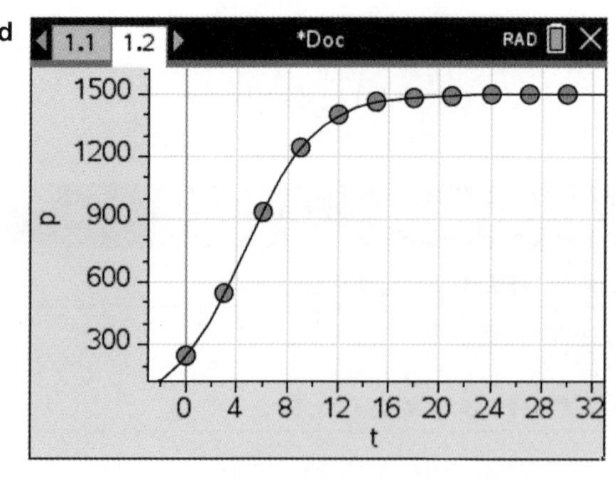

The graph starts at 250 (the initial number of rabbits). This increases exponentially to approach a limiting value (asymptote) of 1500, which is the carrying capacity.

MATCHED EXAMPLE 14

$y = Ae^{\frac{2}{3}x^3} - \frac{1}{2}$

MATCHED EXAMPLE 15

It will take about 9 hours for the number of organisms to increase to 3000.

MATCHED EXAMPLE 16

a

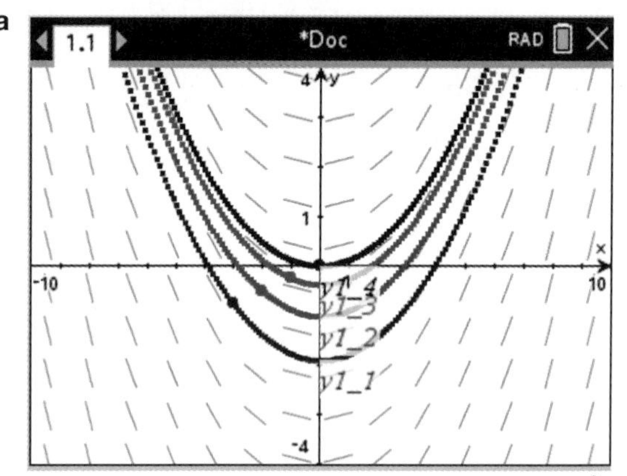

b The curves appear to be a parabola of the form $y = \frac{x^2}{8} + c$.

MATCHED EXAMPLE 17

$(x_1, y_1) = (0.2, 2)$

$(x_2, y_2) = (0.4, 2.008)$

$(x_3, y_3) = (0.6, 2.04)$

MATCHED EXAMPLE 18

a $y = \frac{x^4}{4} + 1$

b $(x_1, y_1) = (0.4, 1)$

$(x_2, y_2) = (0.8, 1.0256)$

$(x_3, y_3) = (1.2, 1.2304)$

$(x_4, y_4) = (1.6, 1.9216)$

$(x_5, y_5) = (2, 3.560)$

c 29%

CHAPTER 10

MATCHED EXAMPLE 1

a

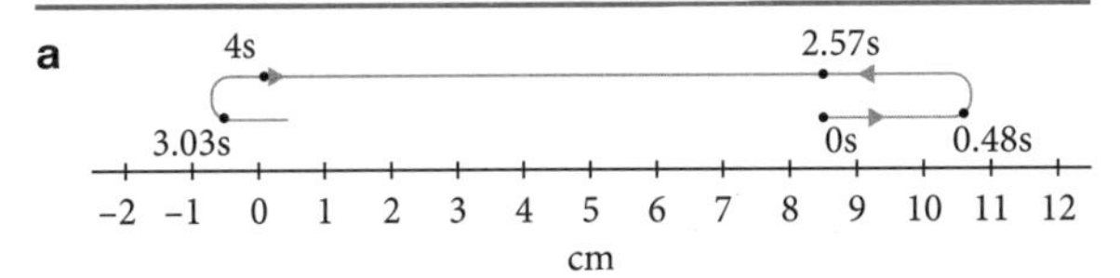

b Average speed = 3.43 cm/s

c Average velocity = −2.16 cm/s

MATCHED EXAMPLE 2

a Average speed = 441.95 km/h

b Average velocity = 411.98 km/h

c The velocity vector has direction 075.16°T.

MATCHED EXAMPLE 3

a The velocity of the particle will be zero at $t = 2$ seconds and $t = 5$ seconds.

When $t = 2$, $x = 17\frac{1}{3}$ cm.

When $t = 5$, $x = 8\frac{1}{3}$ cm.

b $t = 3\frac{1}{2}$, $\frac{dx}{dt} = -4\frac{1}{2}$ m/s

c Total distance is $25\frac{2}{3}$ cm.

MATCHED EXAMPLE 4

a $v = 6t^2 - 4t - 2$

b $x = 2t^3 - 2t^2 - 2t + 1$

c $a = 8$ m/s^2

MATCHED EXAMPLE 5

$$x = 1 - \frac{\sqrt{2}}{\sqrt{3t + \sqrt{2}}}$$

MATCHED EXAMPLE 6

a $k = -5$

b $x = -\frac{1}{5}\log_e(1 - 5t), t \neq \frac{1}{5}$

MATCHED EXAMPLE 7

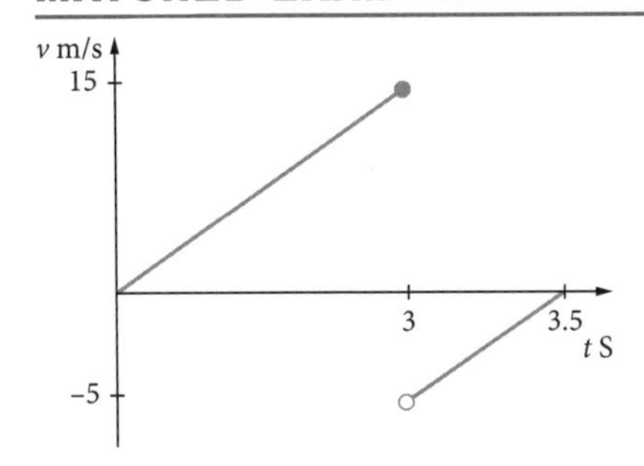

MATCHED EXAMPLE 8

a

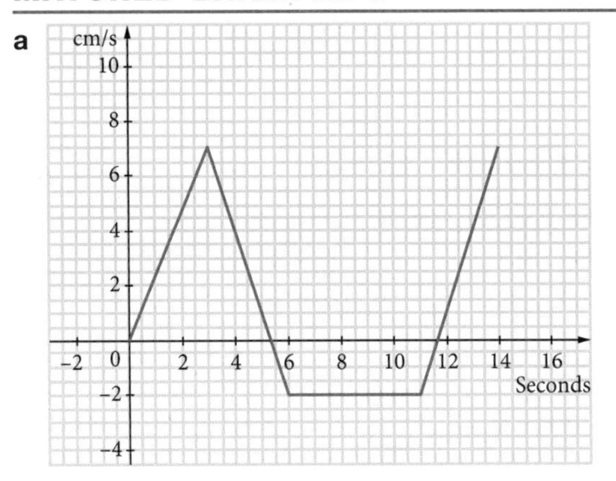

b The car's total distance was 38.17 cm.

c Displacement is 15.51 cm.

MATCHED EXAMPLE 9

a

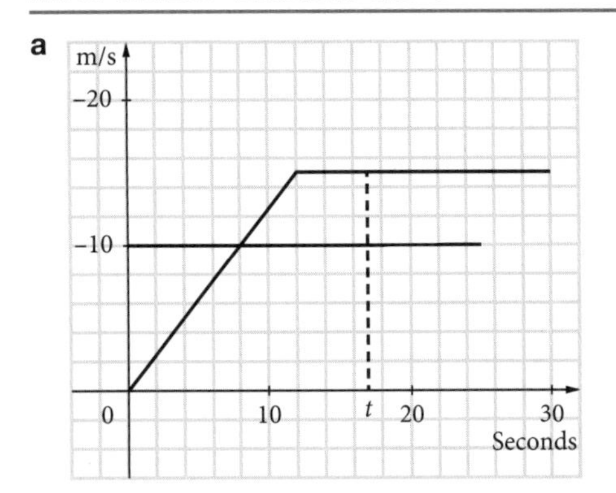

b The van and the electric bike reach the church after 18 seconds.

c 180 m

MATCHED EXAMPLE 10

a $v = \frac{3g}{2k}(1 - e^{-2kt})$

b $k = 0.29$

c The terminal velocity is 51 ms^{-1}.

d The particle has fallen 240 m.

MATCHED EXAMPLE 11

a

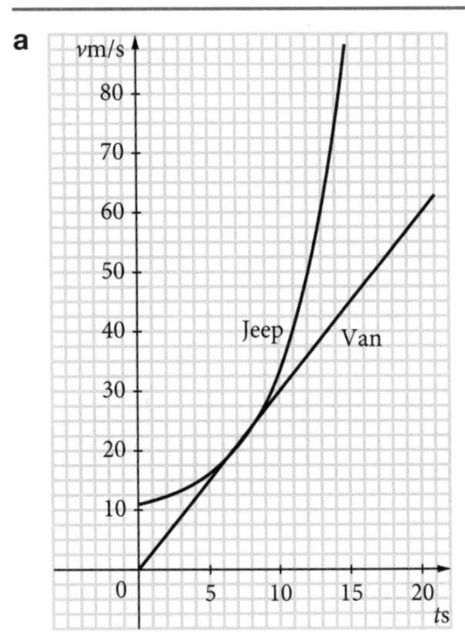

b The jeep passes the van about 20.53 seconds after entering the straight about 662 m down the straight.

MATCHED EXAMPLE 12

The vehicle covers $712\frac{1}{2}+800 = 1512\frac{1}{2}$ m until it comes to rest.

MATCHED EXAMPLE 13

a The Atom Ariel 4 caught up with the Drakan Spyder after 32 seconds.

b The distance from start to finish was 4450 m.

MATCHED EXAMPLE 14

$a=-10, u=20, s=10$

Use $s=ut+\frac{1}{2}at^2$ and solve for t.

$$10=20t+\frac{1}{2}(-10)t^2$$

$$t^2-4t+2=0$$

Using CAS, completing the square, or the quadratic formula gives $2-\sqrt{2}$ as the smaller value of the two solutions for t. This represents the time taken on the downward journey.

The second solution is the time taken for the pendulum to go halfway on its upward journey.

MATCHED EXAMPLE 15

Total time is 11.59 seconds.

MATCHED EXAMPLE 16

a $\frac{dv}{dt}=g-kv$

b $v=\frac{1}{k}(g-(g-k)e^{-kt})$ where $k=0.0937...$

c As t increases, the velocity approaches $\frac{g}{k}=105$ m/s.

d 640 m

MATCHED EXAMPLE 17

$v=227.5$ m/s.

The van has travelled $2666\frac{2}{3}$ m.

MATCHED EXAMPLE 18

a

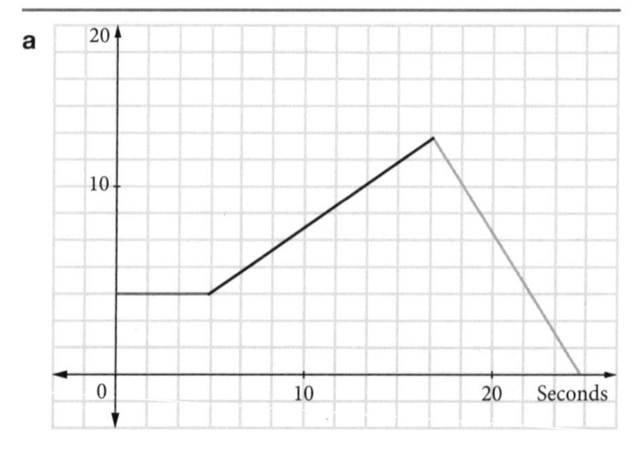

b

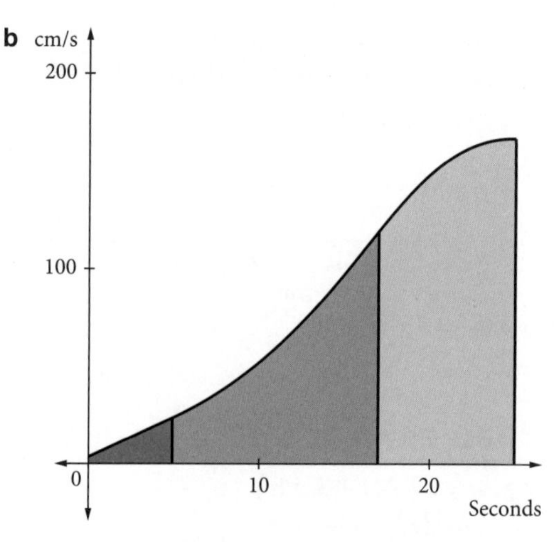

c The distance travelled by the particle is 2029 cm.

CHAPTER 11

MATCHED EXAMPLE 1

a $\underset{\sim}{r}(0)=\underset{\sim}{i}+0\underset{\sim}{j}$

$\underset{\sim}{r}(1)=2\underset{\sim}{i}+2\underset{\sim}{j}$

$\underset{\sim}{r}(-1)=2\underset{\sim}{i}-2\underset{\sim}{j}$

b $\sqrt{41}$

MATCHED EXAMPLE 2

a Cartesian equation: $y=-\frac{2x^2}{25}$

b Domain $x\geq 0$ and range $y\leq 0$

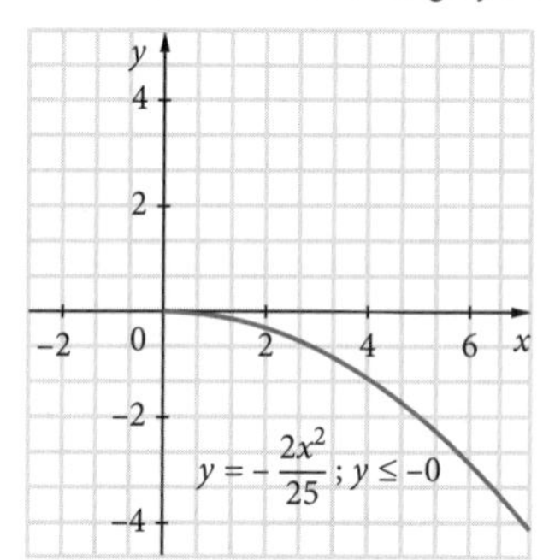

MATCHED EXAMPLE 3

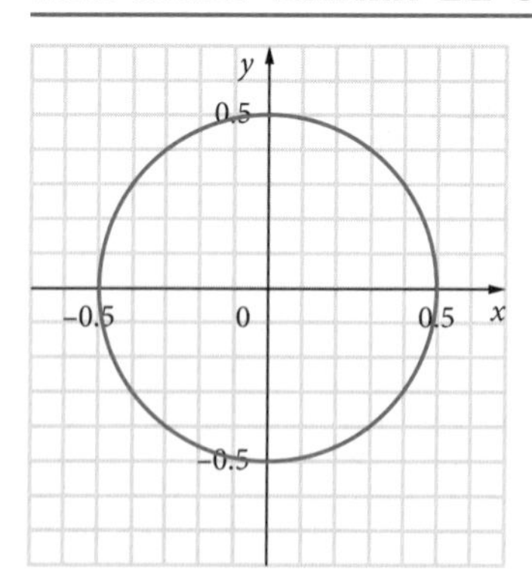

MATCHED EXAMPLE 4

a Velocity, $\dot{\underset{\sim}{r}}(t) = 2t\underset{\sim}{i} - \sin(t)\underset{\sim}{j} + 2e^{2t}\underset{\sim}{k}$

Acceleration, $\ddot{\underset{\sim}{r}}(t) = 2\underset{\sim}{i} - \cos(t)\underset{\sim}{j} + 4e^{2t}\underset{\sim}{k}$

b Velocity, $\dot{\underset{\sim}{r}}\left(\frac{\pi}{2}\right) = \pi\underset{\sim}{i} - \underset{\sim}{j} + 2e^{\pi}\underset{\sim}{k}$

Acceleration, $\ddot{\underset{\sim}{r}}\left(\frac{\pi}{2}\right) = 2\underset{\sim}{i} + 4e^{\pi}\underset{\sim}{k}$

MATCHED EXAMPLE 5

a $|\dot{\underset{\sim}{r}}(t)| = \sqrt{9t^2 + \sin^2(t) + 4e^{2t}}$

b $|\dot{\underset{\sim}{r}}(0)| = 2$

MATCHED EXAMPLE 6

Velocity, $\dot{\underset{\sim}{r}}(t) = t^2\underset{\sim}{i} + (3 - \cos(t))\underset{\sim}{j} + \frac{1}{2}\sin(2t)\underset{\sim}{k}$

Displacement, $\underset{\sim}{r}(t) = \frac{t^3}{3}\underset{\sim}{i} + (3t - \sin(t))\underset{\sim}{j} + \left(\frac{13}{4} - \frac{1}{4}\cos(2t)\right)\underset{\sim}{k}$

MATCHED EXAMPLE 7

a The particles collide at the point (6, 9).

b The particles' paths cross at the points (6, 9) and (–4, 4).

They cross at (–4, 4).

They cross and collide at (6, 9).

MATCHED EXAMPLE 8

a $x^2 + y^2 = 25$

b

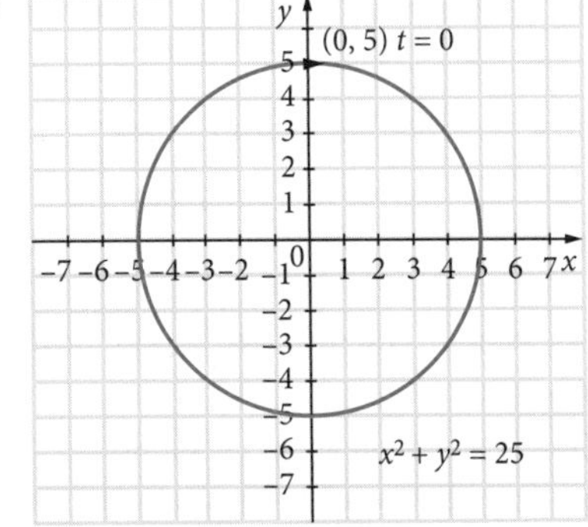

MATCHED EXAMPLE 9

a The time taken to return to the ground is 11.43 s

b The maximum height reached is 160.00 m.

c The initial speed of the object is 59.46 m/s.

CHAPTER 12

MATCHED EXAMPLE 1

a $E(X) = 1.5$

b 7

MATCHED EXAMPLE 2

a 1

b 4

MATCHED EXAMPLE 3

$\frac{9}{5}$

MATCHED EXAMPLE 4

Mean: 11 hours

Variance: 6 hours

MATCHED EXAMPLE 5

Mean: 390

Variance: 378

MATCHED EXAMPLE 6

Mean mass: 4400 g

Standard deviation: $10\sqrt{1711}$

MATCHED EXAMPLE 7

a $E(X + Y) = 132$ and $Var(X + Y) = 25$

b 0.788

MATCHED EXAMPLE 8

a Mean: 1500

SD: $6\sqrt{3}$

b 0.680

c $x = 1479$ mL

MATCHED EXAMPLE 9

$\frac{13}{100}$

MATCHED EXAMPLE 10

a

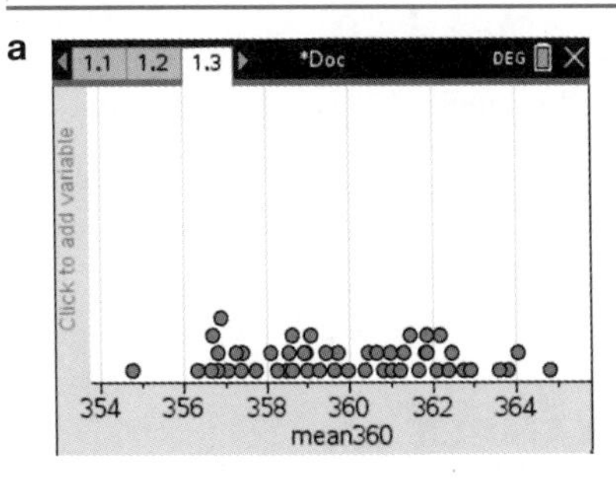

ANSWERS

Or

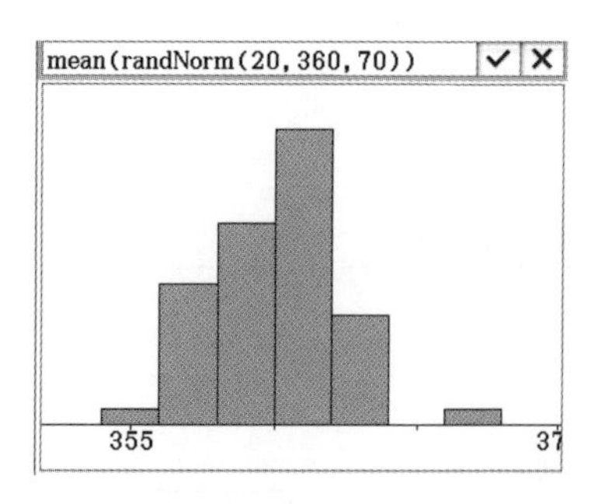

b Mean of $\bar{x} = 359.851$ or Mean of $\bar{x} = 359.877$

MATCHED EXAMPLE 11

Sample size	Mean of $\bar{x}$	Standard deviation of $\bar{x}$
20	120.133	1.7006
100	119.845	0.9021

MATCHED EXAMPLE 12

0.447

MATCHED EXAMPLE 13

(26.39, 63.61)

MATCHED EXAMPLE 14

$\Pr(\bar{X} < 3800) = 0.9987$

MATCHED EXAMPLE 15

95% confidence interval = (82.52, 87.48)

MATCHED EXAMPLE 16

(449.723, 450.277)

MATCHED EXAMPLE 17

$n = 16$

MATCHED EXAMPLE 18

The sample mean of the sample of 49 croissants is 45 g.

MATCHED EXAMPLE 19

H_0: $\mu = 100$

H_1: $\mu > 100$

MATCHED EXAMPLE 20

H_0: $\mu = 80$

H_1: $\mu \neq 80$

MATCHED EXAMPLE 21

The average shelf life of peanut butter is less than what the company claims, 85 days.

($z = -3.098$, $p = 0.00097$, $\bar{x} = 83$, $\sigma = 5$, $n = 60$)

There is enough evidence to support the claim that the average shelf life of peanut butter is less than 85 days.

MATCHED EXAMPLE 22

The mean volume of oil in the bottle is significantly less than 550 ml

($z = -3.14269$, $p = 0.001674$, $\bar{x} = 546$, $\sigma = 9$, $n = 50$).

There is enough evidence to support the claim that the average volume of oil in the bottle is less than 550 ml.

MATCHED EXAMPLE 23

The test statistic $z = -1$ is not less than the critical z-value $= -1.645$ and lies outside the rejection region.

We fail to reject the null hypothesis $\mu = 350$ g at the 0.05 significance level.

($z = -1$, $\bar{x} = 347$, $\sigma = 15$, $n = 25$)

MATCHED EXAMPLE 24

a The critical z-values are -2.326 and 2.326.

b Reject the null hypothesis if the sample mean is less than 80.930 or greater than 89.071.

MATCHED EXAMPLE 25

a H_0: $\mu = 15$ min

The new version of the drone has 15 minute battery life.

H_1: $\mu > 15$ min

The new version of the drone has increased battery life.

b The null hypothesis H_0: $\mu = 15$ min is rejected when it is true. This would mean that you conclude that the new version of the drone has increased battery life when it really does not. Consumers will therefore be wasting their money on the product.

c A type II error occurs if the null hypothesis is not rejected when it is false. This would mean that you conclude that the new version of the drone does not have increased battery life when in fact it has. Therefore, consumers are missing out on an improved product.